Advances in Experimental Medicine and Biology

Volume 1290

Advances in Experimental Medicine and Biology provides a platform for scientific contributions in the main disciplines of the biomedicine and the life sciences. This series publishes thematic volumes on contemporary research in the areas of microbiology, immunology, neurosciences, biochemistry, biomedical engineering, genetics, physiology, and cancer research. Covering emerging topics and techniques in basic and clinical science, it brings together clinicians and researchers from various fields.

Advances in Experimental Medicine and Biology has been publishing exceptional works in the field for over 40 years, and is indexed in SCOPUS, Medline (PubMed), Journal Citation Reports/Science Edition, Science Citation Index Expanded (SciSearch, Web of Science), EMBASE, BIOSIS, Reaxys, EMBiology, the Chemical Abstracts Service (CAS), and Pathway Studio.

2019 Impact Factor: 2.450 5 Year Impact Factor: 2.324

More information about this series at http://www.springer.com/series/5584

Alexander Birbrair

Editor

Tumor Microenvironment

The Role of Interleukins – Part B

 Springer

Editor
Alexander Birbrair
Department of Radiology
Columbia University Medical Center
New York, NY, USA

Department of Pathology
Federal University of Minas Gerais
Belo Horizonte, MG, Brazil

ISSN 0065-2598   ISSN 2214-8019   (electronic)
Advances in Experimental Medicine and Biology
ISBN 978-3-030-55619-8   ISBN 978-3-030-55617-4   (eBook)
https://doi.org/10.1007/978-3-030-55617-4

This Springer imprint is published by the registered company Springer Nature Switzerland AG
The registered company address is: Gewerbestrasse 11, 6330 Cham, Switzerland

This book is dedicated to my mother, Marina Sobolevsky, of blessed memory, who passed away during the creation of this volume. Professor of Mathematics at the State University of Ceará (UECE), she was loved by her colleagues and students, whom she inspired by her unique manner of teaching. All success in my career and personal life I owe to her.

My beloved mom Marina Sobolevsky of blessed memory (July 28, 1959–June 3, 2020)

Preface

This book's initial title was "Tumor Microenvironment". However, due to the current great interest in this topic, we were able to assemble more chapters than would fit in one book, covering tumor microenvironment biology from different perspectives. Therefore, the book was subdivided into several volumes.

This book *Interleukins in the Tumor Microenvironment—Part B* presents contributions by expert researchers and clinicians in the multidisciplinary areas of medical and biological research. The chapters provide timely detailed overviews of recent advances in the field. This book describes the major contributions of different interleukins in the tumor microenvironment during cancer development. Further insights into these mechanisms will have important implications for our understanding of cancer initiation, development, and progression. The authors focus on the modern methodologies and the leading-edge concepts in the field of cancer biology. In recent years, remarkable progress has been made in the identification and characterization of different components of the tumor microenvironment in several tissues using state-of-the-art techniques. These advantages facilitated the identification of key targets and definition of the molecular basis of cancer progression within different organs. Thus, the present book is an attempt to describe the most recent developments in the area of tumor biology, which is one of the emergent hot topics in the field of molecular and cellular biology today. Here, we present a selected collection of detailed chapters on what we know so far about the interleukins in the tumor microenvironment in various tissues. Eight chapters written by experts in the field summarize the present knowledge about distinct interleukins during tumor development.

Zoran Culig from Medical University of Innsbruck describes interleukin-6 function and targeting in prostate cancer. Małgorzata Krzystek-Korpacka and colleagues from Wroclaw Medical University discuss interleukin-7 signaling in the tumor microenvironment. Ramesh B. Batchu and colleagues from Wayne State University School of Medicine address the importance of interleukin-10 signaling in the tumor microenvironment of epithelial ovarian cancer. Gabriella Campadelli-Fiume and colleagues from the University of Bologna compile our understanding of targeted delivery of interleukin-12 adjuvants by oncolytic viruses. Runqiu Jiang and Beicheng Sun from Nanjing University Medical School update us with what we know about interleukin-22 signaling in the tumor microenvironment. Craig A. Elmets and colleagues from the University of Alabama at Birmingham summarize the current knowledge on interleukin-23 in the tumor microenvironment. Rajagopal Ramesh and colleagues from the University of Oklahoma Health Sciences Center focus on how interleukin-24 reconfigures the tumor microenvironment for eliciting antitumor responses. Finally, Alain H. Rook and colleagues from the University of Pennsylvania give an overview of interleukin-31 and its role in the tumor microenvironment.

It is hoped that the articles published in this book will become a source of reference and inspiration for future research ideas. I would like to express my deep gratitude to my wife Veranika Ushakova and Mr. Murugesan Tamilsevan from Springer, who helped at every step of the execution of this project.

Belo Horizonte, Brazil   Alexander Birbrair

Contents

1 Interleukin-6 Function and Targeting in Prostate Cancer 1
   Zoran Culig

2 Interleukin (IL)-7 Signaling in the Tumor
   Microenvironment. 9
   Iwona Bednarz-Misa, Mariusz A. Bromke,
   and Małgorzata Krzystek-Korpacka

3 IL-10 Signaling in the Tumor Microenvironment
   of Ovarian Cancer. 51
   Ramesh B. Batchu, Oksana V. Gruzdyn, Bala K. Kolli,
   Rajesh Dachepalli, Prem S. Umar, Sameer K. Rai,
   Namrata Singh, Pavan S. Tavva, Donald W. Weaver,
   and Scott A. Gruber

4 Targeted Delivery of IL-12 Adjuvants Immunotherapy
   by Oncolytic Viruses. 67
   Andrea Vannini, Valerio Leoni,
   and Gabriella Campadelli-Fiume

5 IL-22 Signaling in the Tumor Microenvironment 81
   Runqiu Jiang and Beicheng Sun

6 IL-23 and the Tumor Microenvironment 89
   Sweta Subhadarshani, Nabiha Yusuf, and Craig A. Elmets

7 Interleukin (IL)-24: Reconfiguring the Tumor
   Microenvironment for Eliciting Antitumor Response 99
   Rajagopal Ramesh, Rebaz Ahmed, and Anupama Munshi

8 Interleukin-31, a Potent Pruritus-Inducing Cytokine
   and Its Role in Inflammatory Disease
   and in the Tumor Microenvironment 111
   Alain H. Rook, Kathryn A. Rook, and Daniel J. Lewis

Index. 129

Contributors

Rebaz Ahmed Department of Pathology, University of Oklahoma Health Sciences Center, Oklahoma City, OK, USA

Graduate Program in Biomedical Sciences, University of Oklahoma Health Sciences Center, Oklahoma City, OK, USA

Ramesh B. Batchu Wayne State University School of Medicine, Detroit, MI, USA

John D. Dingell VA Medical Center, Detroit, MI, USA

Iwona Bednarz-Misa Department of Medical Biochemistry, Wroclaw Medical University, Wroclaw, Poland

Mariusz A. Bromke Department of Medical Biochemistry, Wroclaw Medical University, Wroclaw, Poland

Gabriella Campadelli-Fiume Department of Experimental, Diagnostic and Specialty Medicine, University of Bologna, Bologna, Italy

Zoran Culig Experimental Urology, Department of Urology, Medical University of Innsbruck, Innsbruck, Austria

Rajesh Dachepalli Mcd Manor Organics Pvt. Ltd., Hyderabad, India

Craig A. Elmets Department of Dermatology, O'Neal Comprehensive Cancer Center, University of Alabama at Birmingham, Birmingham, AL, USA

Scott A. Gruber Wayne State University School of Medicine, Detroit, MI, USA

John D. Dingell VA Medical Center, Detroit, MI, USA

Oksana V. Gruzdyn Wayne State University School of Medicine, Detroit, MI, USA

John D. Dingell VA Medical Center, Detroit, MI, USA

Runqiu Jiang Department of Hepatobiliary Surgery, The Affiliated Drum Tower Hospital of Nanjing University Medical School, Nanjing, People's Republic of China

Medical School of Nanjing University, Nanjing, People's Republic of China

Bala K. Kolli Wayne State University School of Medicine, Detroit, MI, USA

John D. Dingell VA Medical Center, Detroit, MI, USA

Med Manor Organics Pvt. Ltd., Hyderabad, India

Małgorzata Krzystek-Korpacka Department of Medical Biochemistry, Wroclaw Medical University, Wroclaw, Poland

Valerio Leoni Department of Experimental, Diagnostic and Specialty Medicine, University of Bologna, Bologna, Italy

Daniel J. Lewis Department of Dermatology, Perelman School of Medicine, University of Pennsylvania, Philadelphia, PA, USA

Anupama Munshi Stephenson Cancer Center, University of Oklahoma Health Sciences Center, Oklahoma City, OK, USA

Department of Radiation Oncology, University of Oklahoma Health Sciences Center, Oklahoma City, OK, USA

Sameer K. Rai Med Manor Organics Pvt. Ltd., Hyderabad, India

Rajagopal Ramesh Department of Pathology, University of Oklahoma Health Sciences Center, Oklahoma City, OK, USA

Stephenson Cancer Center, University of Oklahoma Health Sciences Center, Oklahoma City, OK, USA

Graduate Program in Biomedical Sciences, University of Oklahoma Health Sciences Center, Oklahoma City, OK, USA

Alain H. Rook Department of Dermatology, Perelman School of Medicine, University of Pennsylvania, Philadelphia, PA, USA

Kathryn A. Rook Clinical Dermatology and Allergy, University of Pennsylvania School of Veterinary Medicine, Philadelphia, PA, USA

Namrata Singh Med Manor Organics Pvt. Ltd., Hyderabad, India

Sweta Subhadarshani

Beicheng Sun Department of Hepatobiliary Surgery, The Affiliated Drum Tower Hospital of Nanjing University Medical School, Nanjing, People's Republic of China

Pavan S. Tavva Med Manor Organics Pvt. Ltd., Hyderabad, India

Prem S. Umar Med Manor Organics Pvt. Ltd., Hyderabad, India

Andrea Vannini Department of Experimental, Diagnostic and Specialty Medicine, University of Bologna, Bologna, Italy

Donald W. Weaver Wayne State University School of Medicine, Detroit, MI, USA

Nabiha Yusuf

Interleukin-6 Function and Targeting in Prostate Cancer

1

Zoran Culig

Abstract

Interleukin-6 (IL-6) is a proinflammatory cytokine, which is involved in pathogenesis of several cancers. Its expression and function in prostate cancer have been extensively studied in cellular models and clinical specimens. High levels of IL-6 were detected in conditioned media from cells which do not respond to androgens. Increased phosphorylation of signal transducer and activator of transcription (STAT)3 factor which is induced in response to IL-6 is one of the typical features of prostate cancer. However, reports in the literature show regulation of neuroendocrine phenotype by IL-6. Effects of IL-6 on stimulation of proliferation, migration, and invasion lead to the establishment of experimental and clinical approaches to target IL-6. In prostate cancer, anti-IL-6 antibodies were demonstrated to inhibit growth in vitro and in vivo. Clinically, application of anti-IL-6 therapies did not improve survival of patients with metastatic prostate cancer. However, clinical trial design in the future may include different treatment schedule and combinations with experimental and clinical therapies. Endogenous inhibitors of IL-6 such as suppressors of cytokine signaling and protein inhibitors of activated STAT have variable effects on prostate cells, depending on presence or absence of the androgen receptor.

Keywords

Interleukin-6 · Prostate cancer · Tumor progression · Androgen receptor · Proliferation · Neuroendocrine phenotype · JANUS kinase · STAT factors · Stem cells · Epithelial to mesenchymal transition · Sensitivity · Anti-interleukin-6 antibodies · Galiellalactone · Clinical studies · Endogenous inhibitors of cytokine signaling

1.1 Factors That Regulate Interleukin-6 Expression in Prostate Cancer

Therapy approaches for non-localized prostate cancer (PCa) are based on inhibition of androgenic stimulation of PCa cell growth. In addition to inhibitors of androgen synthesis, blockers of androgen receptor (AR) are available. Previously, most clinical studies were focused on hydroxyflutamide and bicalutamide. Molecular biology studies have revealed that, during continued

Z. Culig (✉)
Experimental Urology, Department of Urology, Medical University of Innsbruck, Innsbruck, Austria
e-mail: zoran.culig@i-med.ac.at

© The Editor(s) (if applicable) and The Author(s), under exclusive license to Springer Nature Switzerland AG 2021
A. Birbrair (ed.), *Tumor Microenvironment*, Advances in Experimental Medicine and Biology 1290, https://doi.org/10.1007/978-3-030-55617-4_1

treatment with antiandrogens, specific mutations in the ligand-binding domain of the AR may occur. Further improvement of antiandrogen therapy in PCa has been achieved with the next-generation drug, enzalutamide. However, there is substantial evidence that tumor progression also occurs in the presence of enzalutamide. One of the reasons may be AR mutations [1]. In addition, constitutively active AR have been described in conditions of therapy resistance [2].

Thus, inhibition of the androgen signaling pathway is not sufficient to prevent PCa progression. Novel targets have been described in tumors, in particular in clinical specimens obtained from metastatic lesions or in patient-derived xenografts. There is an increasing number of studies on interleukins (IL) and chemokines, which regulate proliferation, apoptosis, migration, and invasion in tumor cells. There is a particular interest in the proinflammatory cytokine IL-6, which influences carcinogenesis and progression by regulation of immune response and intracellular signaling. Twille and associates were the first who measured elevated IL-6 levels in supernatants of AR-negative cell lines, whereas IL-6 levels were not detectable in conditioned media from androgen-sensitive LNCaP cells [3]. One explanation for the suppression of IL-6 expression in PCa cells is negative effect of an androgen on nuclear factor (NF) kappa B, which is the main upstream regulator of IL-6 [4]. In addition to NF kappa B, transforming growth factor-beta and members of the activation protein-1 complex contribute to increased expression of IL-6 [5, 6]. In contrast, vitamin D decreases IL-6 expression in prostate cells [7]. It should be mentioned that vitamin D decreases PCa growth by multiple mechanisms including negative regulation of IL-6.

Immunohistochemical analyses in human tissues revealed expression of IL-6 and its receptor in the majority of PCa tissues [8]. In general, it is believed that both autocrine and paracrine loops of IL-6/IL-6 receptor exist in PCa although some of the methodological approaches in morphology studies have been a subject of discussion [9] (Fig. 1.1). High levels of intracellular IL-6 have been measured in tissues from patients with localized PCa [10]. This finding may indicate oncogenic function of IL-6 during early stages of prostate carcinogenesis.

1.2 Interleukin-6 Activation of Signaling Pathways in Prostate Tumors

IL-6 receptor consists of the ligand-binding subunit gp80 and the signal-transducing subunit gp130. Alterations of the receptor have not been reported in PCa. Binding of ligand to the receptor leads to activation of Janus kinases and signal transducer and activator of transcription (STAT) factor 3. In addition to STAT3, other STAT factors such as STAT5 have a documented role in PCa progression [11]. STAT5 is an important target in PCa therapy and is known to regulate cellular plasticity.

It was proposed that IL-6 has a role in regulation of cellular processes in pre-neoplastic lesions, such as prostate intraepithelial neoplasia [12]. It is, however, known that there is a limited number of cellular models for prostate intraepithelial neoplasia available and more studies are needed to document the role of IL-6 in precursor lesions. The presence of activated STAT3 is considered a surrogate for malignancy, and STAT3 is targeted in multiple cancers. However, in case of LNCaP cells, activation of STAT3 in response to stromal-epithelial interactions has also been described [13]. The report by Degeorges and associates is therefore consistent with the influence of IL-6 on neuroendocrine differentiation in PCa [13, 14]. IL-6 may thus cause a growth arrest in a subgroup of prostate tumors as evidenced by upregulation of the cell cycle inhibitor p27 [15]. STAT3 may act in a subgroup of PCa patients as a tumor suppressor by inducing cellular senescence through the ARF – Mdm2 p53 tumor suppressor axis [16]. Further studies to determine the clinical significance of those findings in human PCa at different stages are needed.

In contrast to earlier studies in which the importance of neuroendocrine differentiation in clinical PCa was not recognized, recent research

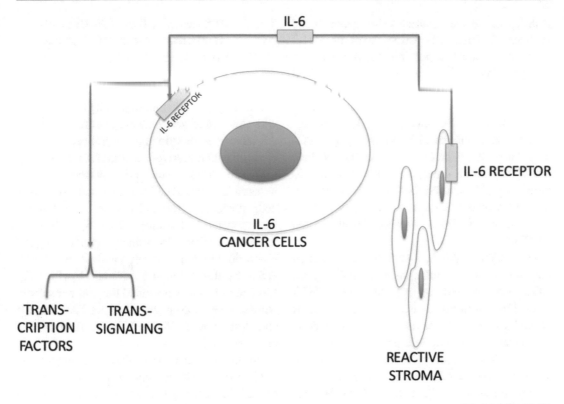

IL-6

IL-6 RECEPTOR

IL-6 RECEPTOR

IL-6
CANCER CELLS

TRANS-
CRIPTION
FACTORS

TRANS-
SIGNALING

REACTIVE
STROMA

RS© by Prof. Dr. Zoran Culig 2020

Fig. 1.1 Cancer cells, benign cells, and adjacent stromal cells may be a source of IL-6. Multiple autocrine and paracrine loops are described in the literature

highlighted the importance of neuroendocrine phenotype in different subgroups of PCa. The mechanism of IL-6 regulation of the neuroendocrine phenotype includes the Etk/Bmx kinase which is downstream to phosphatidylinositol 3-kinase [17]. A constitutively active subunit of phosphatidylinositol 3-kinase (p110) induced neuroendocrine phenotype in absence of IL-6. Neuroendocrine cells themselves influence the proliferation of adjacent epithelial cells by secretion of peptides that bind to specific receptors. In PCa cells, IL-6 was shown to induce the phosphatidylinositol 3-kinase pathway, thus inhibiting apoptosis [18]. Induction of neuroendocrine phenotype by IL-6 is associated with suppression of RE-1 transcription factor (REST) [19]. One has to keep in mind that neuroendocrine differentiation is regulated by multiple factors, mostly by second messengers and/or androgen ablation.

Evidence for oncogenic effects of STAT3 in PCa was obtained in several studies from the Gao laboratory. In some experimental approaches, the use of the signal transduction inhibitor AG 490 with subsequent downregulation of phosphorylated STAT3 caused the growth-inhibitory effect [20]. Even more, overexpression of STAT3 in PCa cells promotes androgen-independent growth of cells, which express the AR [21]. STAT3 activation may be achieved by other growth factors such as epidermal growth factor or IL-11 [22].

Studies on IL-6 signaling in PCa are also focused on mitogen-activated protein kinase (MAPK), which are involved in the proliferative effect of IL-6 [23]. It is important to mention that, in cells treated with IL-6, the ErbB2 oncogene associates with the IL-6 receptor. These findings may provide some important explanations for regulation of proliferative responses and signaling pathways in different prostate cellular models and sublines. Different expression levels of ErbB2 may be therefore responsible for a weaker or stronger effect of IL-6 in a PCa subline derived during tumor progression.

Clinically, activated STAT3 is observed in more than 80% of PCa cases [24]. STAT3 expression was also confirmed in the majority of PCa metastases [25]. Therefore, one could conclude that STAT3 is an oncogene although in a subgroup of patients, it may act as a tumor suppressor. Biomarkers are needed to identify patients who will benefit from targeting phosphorylated STAT3.

IL-6-related cytokines such as oncostatin M may also regulate cellular processes in PCa. Such effect was reported for oncostatin M in PC-3 cells [26]. Additional experimental work is needed to clarify whether there is an upregulation of oncostatin M during PCa progression.

IL-6 effect on epithelial to mesenchymal transition is important for better understanding of therapy resistance [27]. It was documented that IL-6 effects during epithelial to mesenchymal transition are mediated by heat-shock protein 27. Heat-shock protein 27 is therefore recognized as a chaperone, which is a target in advanced PCa.

STAT3 activation is also of crucial importance for the generation of stem cells, which could not be targeted by conventional therapies [28]. These most primitive cells require STAT3 for survival and clonogenic ability. The process of generation of stem cells is supported by enhanced production of reactive oxygen species [29]. Loss of AR expression is also causally related to generation of stem cells [30].

There is an interplay between signaling of IL-6 and interferon regulatory factor 9 (IRF9) in PCa [31]. The expression of IRF9 correlates with that of IL-6. On the other hand, IRF9 sensitizes the cells to the antiproliferative effect of interferon alpha 2.

1.3 Interleukin-6 and Androgen Responsiveness in Prostate Cancer

Nonsteroidal activation of the AR may be important in PCa progression, especially in conditions with low androgen concentrations. Functional AR have been described in several cellular models and patient-derived xenografts representing advanced PCa. For this reason, ligand-independent and synergistic AR activation have been in a focus of research interest. AR activation by a nonsteroidal compound (IL-6) is different from that achieved by a ligand. Importantly, the N-terminal region is mostly involved in AR activation by IL-6 through STAT3 and MAPK [32]. Lin and colleagues extended the initial results in which it was shown that IL-6 activates the AR and described positive effects of IL-6 on AR expression [33].

As mentioned before, STAT3 is involved in regulation of AR activity. It was also demonstrated that androgenic hormones may have an effect on expression of STAT3 target genes [34].

In recent years, many studies have addressed the role of different coactivators in the process of steroid receptor activation. P300 and SRC-1 are involved in regulation of functional activity of multiple steroid receptors, including the AR. Both coactivators are necessary for the process of AR activation by IL-6, and their downregulation impairs the effect of IL-6 [35, 36]. In this context, targeting coactivators is considered a specific therapy option in metastatic PCa. This seems to be justified because of coactivator pro-tumorigenic activity and involvement in resistance to endocrine therapies and chemotherapies in cancer. SRC-1 is also phosphorylated in response to IL-6 treatment of PCa cells. The involvement of the N-terminal of the AR in nonsteroidal activation opens the possibility to design more effective treatments targeting the variable N-terminal domain of the AR.

1.4   Changes in Sensitivity to Interleukin-6 During Prostate Cancer Progression

PCa is a slow-growing neoplasm and the exposure of tumor cells to IL-6 may be investigated during a long-term period. Therefore, it was reasonable to study IL-6 intracellular signaling at different stages of carcinogenesis. LNCaP cells are a good model for this type of experimental studies because they were initially inhibited by IL-6. We have observed a decrease of STAT3 phosphorylation along with lack of inhibitory response to IL-6 during tumor progression [37]. Changes in cellular response to IL-6 were confirmed by Ge and colleagues [38]. The authors have observed that regulation of neuroendocrine differentiation by IL-6 is abolished during prolonged treatment with the cytokine.

1.5   Interleukin-6 and Prostate Cancer Therapy

On the basis of most of the experimental studies mentioned in this chapter, one could conclude that options for IL-6 inhibition should be explored. During many years, preclinical studies provided basis for clinical trials. There is an agreement that the results of preclinical studies imply that subgroups of PCa patients may benefit from this kind of therapy. However, it is also necessary to consider appropriate timing of anti-IL-6 therapy. Initial therapy studies were carried out in the PC-3 model, in which high levels of IL-6 were measured [39]. Antibody (CNTO328, siltuximab) treatment resulted in increased apoptosis in the absence or presence of etoposide. The antibody CNTO328 demonstrated the effect on preventing progression of the LAPC-4 xenograft [40]. This treatment has also anti-oncogenic effect on the coactivators p300 and CBP, which are upregulated after androgen ablation. This therapy also diminishes the expression of Mcl-1, which is highly expressed in cells subjected to androgen withdrawal [41]. Mcl-1 is an important mediator of the antiapoptotic action of IL-6.

Clinical studies with the anti-IL-6 antibody were performed in patients with castration therapy-resistant PCa [42, 43]. So far, there is no evidence that CNTO328 administration yields clinical improvement. Several questions may be asked in that context. One could expect that starting treatment at an earlier time point may affect cellular stemness, thus leading to a more favorable clinical response.

Another possibility to enhance anti-STAT3 therapy in PCa is the use of galiellalactone, which was described in PCa cell lines and explants [44]. This compound also inhibits AR signaling in PCa. Thus, combined anti-AR and anti-STAT3 in advanced PCa may have an impact on multiple target genes and signaling pathways and should be further investigated.

1.6   Endogenous Inhibitors of Interleukin-6 Signaling

Suppressors of cytokine signaling (SOCS) prevent continuous activation of STAT factors. They act as a part of an important feedback mechanism also in nonmalignant conditions (Fig. 1.2). Interference with cytokine signaling is enhanced not only by the presence of SOCS, but also by protein inhibitors of activated STAT (PIAS). In case of SOCS-3, it seems that STAT3-dependent and -independent effects occur in cancer cells. In androgen-insensitive cancer cells, SOCS-3 inhibits programmed cell death [45]. The fact that SOCS-3 may influence proliferation and apoptosis in different directions is most probably responsible for lack of therapeutic SOCS-based therapy applications in prostate oncology. In addition to SOCS-3, SOCS-1 was investigated in PCa. There is also an evidence showing that SOCS-1 causes growth inhibition in PCa through preventing cell cycle progression [46]. PIAS1 was found to be associated with PCa cell survival and chemotherapy resistance [47].

In summary, earlier studies on IL-6 in PCa have considerable translational importance. There is no definitive answer to the question whether there is a therapy option in which IL-6 could be co-targeted in PCa. A more detailed

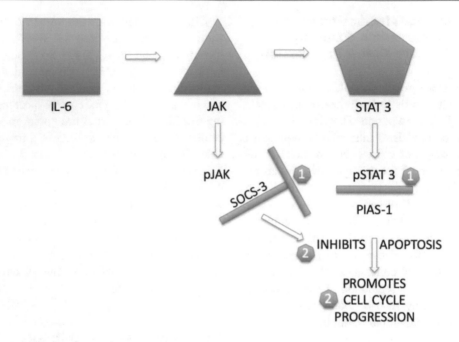

Fig. 1.2 Binding of IL-6 to its receptor leads to activation of the signaling pathway of JAK/STAT3. This activation may be prevented by SOCS and PIAS. However, these proteins also have STAT3-independent effects on proliferation and apoptosis

classification for prostate tumor microenvironment may be helpful in order to characterize the role of IL-6 in stromal-to-epithelial interactions in angiogenesis. For example, the function of IL-6 in pericytes during prostate carcinogenesis may be a subject of major interest [48].

If further studies consider IL-6 impact on cellular stemness and activation of multiple signaling pathways, the anti-IL-6 therapy approach may be revisited.

References

1. Wu Z, Wang K, Yang Z et al (2020) A novel androgen receptor antagonist JJ-450 inhibits enzalutamide-resistant mutant ARF876L nuclear import and function. Prostate 80:319–328
2. Jimenez-Vacas JM, Herrero-Aguayo V, Montero-Hidalgo AJ. et al. Dysregulation of the splicing machinery is directly associated to aggressiveness of prostate cancer. EBioMedicine 51:102547
3. Twillie DA, Eisenberger MA, Carducci MA, Hseih WS, Kim WY, Simons JW (1995) Interleukin-6: a candidate mediator of human prostate cancer morbidity. Urology 45:542–549
4. Keller ET, Chang C, Ershler WB (1996) Inhibition of Nfkappa B activity through maintenance of IkappaBalpha levels contributes to dihydrotestosterone-mediated repression of the interleukin-6 promoter. J Biol Chem 271:26265–26267
5. Park JI, Lee MG, Cho K et al (2003) Transforming growth factor-beta1 activates interleukin-6 expression in prostate cancer through the synergistic collaboration of the Smad2, p38-NF-kappaB, JNK, and Ras signaling pathways. Oncogene 22:4314–4332
6. Zerbini LF, Wang Y, Cho JY, Libermann TA (2003) Constitutive activation of nuclear factor kappaB p50/p65 and Frau-1 and JunD is essential for deregulated interleukin 6 expression in prostate cancer. Cancer Res 63:2206–2215
7. Noon L, Peng L, Feldman D, Peehl DM (2006) Inhibition of p38 by vitamin D reduces interleukin-6 production in normal prostate cells via mitogen-activated protein kinase phosphatase 5: implications for prostate cancer prevention. Cancer Res 66:4516–4524
8. Hobisch A, Rogatsch H, Hittmair A et al (2000) Immunohistochemical localization of interleukin-6 and its receptor in benign, premalignant and malignant prostate tissue. J Pathol 191:239–244
9. Yu SH, Zheng Q, Esopi D et al (2015) A paracrine role for IL6 in prostate cancer patients: lack of production by primary or metastatic tumor cells. Cancer Immunol Res 3:1175–1184

10. Giri D, Ozen M, Ittmann M (2001) Interleukin-6 is an autocrine growth factor in human prostate cancer. Am J Pathol 159:2159–2165

11. Maranto C, Udhane V, Hoang DT et al (2018) JTAT3/MD blockade sensitizes prostate cancer to radiation through inhibition of RAD51 and DNA repair. Clin Cancer Res 24:1917–1931

12. Liu XH, Kirschenbaum A, Liu M et al (2002) Prostaglandin E(2) stimulates prostatic intraepithelial neoplasia cell growth through activation of the interleukin-6/gp130/STAT-3 signaling pathway. Biochem Biophys Res Commun 290:249–255

13. Degeorges A, Tatoud R, Fauvel-Lafeve M et al (1885) Stromal cells from human benign prostate hyperplasia produce a growth-inhibitory factor for LNCaP prostate cancer cells, identified as interleukin-6. Int J Cancer 68:207–214

14. Palmer J, Ernst M, Hammacher A, Hertzog PJ (2005) Constitutive activation of gp130 leads to neuroendocrine differentiation in vitro and in vivo. Prostate 62:282–289

15. Mori S, Murakami-Mori K, Bonavida B (1999) Interleukin-6 induces G1 arrest through induction of p27(Kip1), a cyclin-dependent kinase inhibitor, and neuron-like morphology in LNCaP prostate tumor cells. Biochem Biophys Res Commun 257:809–814

16. Pencik J, Schlederer M, Gruber W et al (2015) STAT3-regulated ARF expression suppresses prostate cancer metastasis. Nat Commun 6:7736

17. Qiu Y, Robinson D, Pretlow TG, Kung HJ (1998) Etk/Bmx, a tyrosine kinase with a pleckstrin-homology domain, is an effector of phosphatidylinositol-3 kinase and is involved in interleukin-6-induced neuroendocrine differentiation of prostate cancer cells. Proc Natl Acad Sci U S A 95:3644–3649

18. Chung TD, Yu JJ, Kong TA, Spiotto MT, Lin JM (2000) Interleukin-6 activates phosphatidylinositol-3-kinase, which inhibits apoptosis in human prostate cancer cell lines. Prostate 42:1–7

19. Zhu Y, Liu C, Cui Y, Nadiminty N, Lou W, Gao AC (2014) Interleukin-6 induces neuroendocrine differentiation (NED) through suppression of silencing transcription factor (REST). Prostate 74:1086–1094

20. Ni Z, Lou W, Leman ES, Gao AC (2000) Inhibition of constitutively activated Stat3 signaling pathway suppresses growth of prostate cancer. Cancer Res 60:1225–1228

21. Lee SO, Lou W, Hou M, de Miguel F, Gerber L, Gao AC (2003) Interleukin-6 promotes androgen-independent growth in LNCaP human prostate cancer cells. Clin Cancer Res 9:370–376

22. Campbell CL, Jiang Z, Savarese DM, Savarese TM (2001) Increased expression of the interleukin-11 receptor and evidence of STAT3 activation in prostate carcinoma. Am J Pathol 158:25–32

23. Qiu Y, Ravi L, Kung HJ (1998) Requirement of ErbB2 for signaling by interleukin-6 in prostate carcinoma cells. Nature 393:83–85

24. Mora L, Buettner R, Seigne J et al (2002) Constitutive activation of Stat3 in human prostate tumors and cell lines: direct inhibition of Stat3 induces apoptosis of prostate cancer cells. Cancer Res 62:6659–6666

25. Don-Doncow N, Marginean F, Loleman I et al (2017) Expression of STAT3 in prostate cancer metastasis. Eur Urol 71:313–316

26. RUscunio H, Domenico R, Ciliberto G, Toniatti C, Travali S, D'Alessandro N (1999) Blocking signaling through the Gp130 chain by interleukin-6 and oncostatin M inhibits PC-3 cell growth and sensitizes the tumor cells to etoposide and cisplatin-mediated cytotoxicity. Cancer 85:124–144

27. Shiota M, Bishop JL, Nip KM et al (2013) Hsp27 regulates epithelial mesenchymal transition, metastasis, and circulating tumor cells in prostate cancer. Cancer Res 73:3109–3119

28. Kroon P, Berry PA, Stower MJ et al (2013) JAK-STAT blockade inhibits tumor initiation and clonogenic recovery of prostate cancer stem-like cells. Cancer Res 73:5288–5298

29. Qu Y, Qyan AM, Liu R et al (2013) Generation of prostate tumor-initiating cells is associated with elevation of reactive oxygen species and IL-6/STAT3 signaling. Cancer Res 73:7090–7100

30. Schroeder A, Herrmann A, Cherryholmes G et al (2014) Loss of androgen receptor expression in stem-like cell phenotype in prostate cancer through STAT3 signaling. Cancer Res 74:1227–1237

31. Erb HHH, Langlechner RV, Moser PL et al (2013) IL6 sensitizes prostate cancer to the antiproliferative effect of IFNalpha2 through IRF9. Endocr Relat Cancer 20:677–689

32. Ueda T, Bruchovsky N, Sadar MD (2002) Activation of the androgen receptor N-terminal domain by interleukin-6 via MAPK and STAT3 signal transduction pathways. J Biol Chem 277:7076–7085

33. Lin DL, Whitney MC, Yao Z, Keller ET (2001) Interleukin-6 induces androgen responsiveness in prostate cancer cells through up-regulation of androgen receptor expression. Clin Cancer Res 7:1773–1781

34. Matsuda T, Junicho A, Yamamoto T et al (2001) Cross-talk between signal transducer and activator of transcription 3 and androgen receptor signaling in prostate carcinoma cells. Biochem Biophys Res Commun 283:179–187

35. Debes JD, Schmidt LJ, Huang H, Tindall DJ (2002) P300 mediates androgen-independent transactivation of the androgen receptor by interleukin 6. Cancer Res 62:5632–5636

36. Ueda T, Mawji NR, Bruchovsky N, Sadar MD (2002) Ligand-independent activation of the androgen receptor by interleukin-6 and the steroid receptor coactivator-1 in prostate cancer cells. J Biol Chem 277:38087–38094

37. Steiner H, Rogatsch H, Berger AP et al (2003) Accelerated in vivo growth of prostate tumors that up-regulate interleukin-6 is associated with reduced retinoblastoma protein expression and activation of the mitogen-activated protein signaling pathway. Am J Pathol 162:655–663

38. Ge D, Gao AC, Zhang Q, Liu S, Xue Y, You Z (2012) LNCaP prostate cancer cells with autocrine interleukin-6 expression are resistant to induced neuroendocrine differentiation due to increased expression of suppressors of cytokine signaling. Prostate 72:1306–1316

39. Smith PC, Keller ET (2001) Anti-interleukin-6 monoclonal antibody induces regression of human prostate cancer xenografts in nude mice. Prostate 48:47–53

40. Wallner L, Dai J, Escara-Wilke J et al (2006) Inhibition of interleukin-6 with CNTO328, an anti-interleukin-6 monoclonal antibody, inhibits conversion to an androgen-independent phenotype in orchiectomized mice. Cancer Res 66:3087–3095

41. Cavarretta IT, Neuwirt H, Untergasser G et al (2007) The antiapoptotic effect of IL-6 autocrine loop in a cellular model of advanced prostate cancer is mediated by Mcl-1. Oncogene 26:2822–2832

42. Dorff T, Goldman B, Pinski J et al (2010) Clinical and correlative results of SWOG S0354: a phase II trial of CNTO 328 (siltuximab), a monoclonal antibody against interleukin-6, in chemotherapy-pretreated patients with castration-resistant prostate cancer. Clin Cancer Res 16:3028–3034

43. Fizazi K, de Bono JS, Flechon A et al (2012) Randomised phase II study of siltuximab (CNTO 328), an anti-IL-6 monoclonal antibody, in combination with mitoxantrone/prednisone versus mitoxantrone/prednisone in metastatic castration-resistant prostate cancer. Eur J Cancer 48:85–93

44. Handle F, Puhr M, Schäfer G et al (2018) The STAT3 inhibitor galiellalactone reduces IL-6-mediated AR activity in benign and malignant prostate models. Mol Cancer Ther 17:2722–2731

45. Puhr M, Santer FR, Neuwirt H et al (2009) Down-regulation of suppressor of cytokine signaling-3 causes prostate cancer cell death through activation of the extrinsic and intrinsic apoptosis pathways. Cancer Res 69:7375–7384

46. Neuwirt H, Puhr M, Santer FR et al (2009) Suppressor of cytokine signaling (SOCS)-1 is expressed in human prostate cancer and exerts growth-inhibitory function through down-regulation of cyclins and cyclin-dependent kinases. Am J Pathol 174:1921–1930

47. Puhr M, Hoefer J, Neuwirt H et al (2014) PIAS1 is a crucial factor for prostate cancer cell survival and a valid target in docetaxel resistant cells. Oncotarget 5:12043–12056

48. Birbrair A, Zhang T, Wang ZM et al (2014) Type-2 pericytes participate in normal and tumoral angiogenesis. Am J Physiol Cell Physiol 307:C25–C38

Interleukin (IL)-7 Signaling in the Tumor Microenvironment

Iwona Bednarz-Misa, Mariusz A. Bromke, and Małgorzata Krzystek-Korpacka

Abstract

Interleukin (IL)-7 plays an important immunoregulatory role in different types of cells. Therefore, it attracts researcher's attention, but despite the fact, many aspects of its modulatory action, as well as other functionalities, are still poorly understood. The review summarizes current knowledge on the interleukin-7 and its signaling cascade in context of cancer development. Moreover, it provides a cancer-type focused description of the involvement of IL-7 in solid tumors, as well as hematological malignancies.

The interleukin has been discovered as a growth factor crucial for the early lymphocyte development and supporting the growth of malignant cells in certain leukemias and lymphomas. Therefore, its targeting has been explored as a treatment modality in hematological malignancies, while the unique ability to expand lymphocyte populations selectively and without hyperinflammation has been used in experimental immunotherapies in patients with lymphopenia. Ever since the early research demonstrated a reduced growth of solid tumors in the presence of IL-7, the interleukin application in boosting up the anticancer immunity has been investigated. However, a growing body of evidence indicative of IL-7 upregulation in carcinomas, facilitating tumor growth and metastasis and aiding drug-resistance, is accumulating. It therefore becomes increasingly apparent that the response to the IL-7 stimulus strongly depends on cell type, their developmental stage, and microenvironmental context. The interleukin exerts its regulatory action mainly through phosphorylation events in JAK/STAT and PI3K/Akt pathways, while the significance of MAPK pathway seems to be limited to solid tumors. Given the unwavering interest in IL-7 application in immunotherapy, a better understanding of interleukin role, source in tumor microenvironment, and signaling pathways, as well as the identification of cells that are likely to respond should be a research priority.

Keywords

Immunotherapy · Immunosurveillance · Lymphangiogenesis · Epithelial-mesenchymal transition (EMT) · Lymphoma · Leukemia · Carcinoma · Metastasis · Signaling pathway · Tumor microenvironment · Apoptosis · Drug resistance

I. Bednarz-Misa · M. A. Bromke
M. Krzystek-Korpacka (✉)
Department of Medical Biochemistry, Wroclaw Medical University, Wroclaw, Poland
e-mail: malgorzata.krzystek-korpacka@umed.wroc.pl

2.1 Introduction

Interleukin (IL)-7 has been discovered as a growth factor crucial for the early lymphocyte development, and thus supporting the growth of malignant cells in certain leukemias and lymphomas. Its targeting has been explored as a treatment modality in hematological malignancies. In turn, its unique ability to expand lymphocyte populations selectively and without hyperinflammation has been employed in experimental immunotherapies in patients with lymphopenia. Early research has shown a reduced growth of solid tumors in the presence of IL-7, accompanied by tumor infiltration with cytotoxic lymphocytes. It has earned the cytokine a place among the "Top Agents with High Potential for Use in Treating Cancer" [1]. However, contradicting an advocated antitumor activity, the available clinical data almost unanimously indicate IL-7 accumulation [2–6] and/or IL-7R overexpression [7–10] in solid tumors.

The excellent reviews on IL-7 biology [11–15] and the interleukin potential as an immunotherapeutic agent [16, 17] are available. The present article focuses on IL-7/IL-7R signaling in tumor milieu. We describe the potential sources of IL-7 in the environment of hematological cancers and solid tumors and present cells displaying the cytokine receptor on their surface. We enumerate factors affecting the expression of IL-7 and its receptor and provide an overview of IL-7 interaction with extracellular matrix and of main signal transduction pathways triggered by the interleukin. Moreover, we review what little is known about IL-7/IL-7R signaling in reference to loose interpretation of capabilities, the acquiring of which determines successful cancer development as described by Hanahan and Weinberg [18]. Furthermore, we present the interleukin signaling from cancer type perspective.

2.2 Interleukin (IL)-7

The active form of human IL-7 is a 25 kD glycoprotein consisting of 177 amino acids forming four α-helixes and a hydrophobic core. Murine interleukin shares 55% homology with human IL-7 and consists of 154 amino acids (reviewed in [11–15]). In addition to the canonical form, a functional alternative splice variant lacking exon 5 (IL-7δ5) has been cloned from human cancer cell lines [19]. The IL-7 can act as a mitogen and a trophic, survival, and differentiation factor for various immune cells, particularly those of the lymphoid lineage. Whether the interleukin is necessary and the type of function it performs depend on the type of cells and their stage of development. It may differ between species as well. As an illustration, IL-7 is essential for T cells in mice and humans while its engagement in the development of B and natural killer (NK) cells is indispensable only in mice. It primarily supports early developmental stages of T and B cells, but it is required again for homeostasis and proper functioning of mature T cells and for promoting survival of mature NK cells (reviewed in [11–15]). In case of the non-lymphoid lineages, IL-7 increases populations of splenic myeloid cells in mice and raises the number of neutrophils and monocytes in circulation. However, the effect is rather indirect due to lack of the IL-7 receptor (IL-7R) on common myeloid progenitors, immature or mature neutrophils, or on the most of cell of myelomonocytic origin, except for bone marrow macrophages [20–22]. The IL-7 promotes the genesis of secondary lymphoid organs [23], lymphatic drainage [24], and lymph node remodeling [25]. Regarding cancer, IL-7 functionality may be beneficial for the host, mostly by eliciting antitumor immune responses, or unfavorable, by supporting the tumor growth. This may happen either directly or by modulating the tumor's microenvironment. Moreover, IL-7 has an oncogenic potential, and transgenic mice spontaneously develop T and B lymphomas [26, 27].

Surprisingly little is known about cellular origin and regulation of IL-7 expression. The IL-7 is considered a tissue-based interleukin, expressed mainly by stromal and epithelial cells, although the precise identity of IL7-secreting cells often remains unknown. In hematological malignancies, potential sources of the interleukin in the thymus are subsets of epithelial cells [28–30] and fibroblasts [31]. In the bone marrow, IL-7 is expressed by sinusoidal endothelial cells, stromal reticular cells [32, 33], mesenchymal stem and progenitor cells (MSCs) [32, 34], and by osteoblasts [35, 36]. In the secondary lymphoid organs, IL-7 is produced by fibroblastic reticular cells, follicular dendritic cells and by endothelial and endothelial cells [37–39], including lymphatic endothelial cells [24]. Interestingly, all the studies employing IL-7 reporter bacterial artificial chromosome (BAC) transgenic mice have uniformly failed to confirm IL-7 expression in the spleen (reviewed in [13]). Lymphocytes T and B do not express IL-7, except for intestinal $\alpha\beta$T cells [40]. In turn, IL-7 may be released by some subsets of dendritic cells [41, 42] and granulocytes [43]. Potential source of the interleukin in solid tumors are fibroblasts [37], vascular endothelial and smooth muscle cells [37, 38, 44], lymphatic endothelial cells [24], epithelial cells [44–47], keratinocytes [48], chondrocytes [49], normal neuronal progenitor cells [50], neurons [51], microglia [52], and astrocytes [52].

The ability to express detectable levels of IL-7 seems to be associated with malignancy. Corroborating the notion, the interleukin has been found in melanoma cells, but not in normal melanocytes [53] or in hepatocellular carcinoma cells, but not in normal hepatocytes [54–56]. Some of the cancer cells co-express IL-7 and its receptor, raising a possibility of an autocrine mode of action. The concomitant IL-7 and IL-7R expression has been reported in Hodgkin lymphoma [57], thyroid lymphoma [38], hepatocellular carcinoma [54], melanoma [53], glioma [4], and in prostate [47], breast [3], and non–small cell lung cancer [6, 58]. Noncancerous cells from tumor microenvironment, including cancer-associated fibroblasts (CAFs) in Hodgkin's lymphoma [57], epithelial cells in lymphoid stroma-rich Warthin's tumor [59], and keratinocytes in patients with cutaneous T-cell lymphoma (CTCL) [60], express IL-7 and its receptor as well. Bone marrow stromal cells may also express functional, albeit truncated, receptor [61, 62]. Moreover, the co-expression of IL-7 and its receptor has been observed in lymphatic [24] and vascular [44] endothelial cells, macrophages [20, 63], chondrocytes [49], and in the intestinal epithelial cells [3, 64].

The synthesis of IL-7 was believed to be constitutive, with the interleukin availability regulated exclusively at the receptor level [65]. However, recent evidence has shown that IL-7 expression in the specific hematopoietic stem cell niches of the bone marrow might be regulated by membrane-type 1 matrix metalloproteinase (MMP-14), facilitating cell fate decisions. The interleukin expression has been induced by hypoxia-inducible factor (HIF)-1α. Noteworthy, MMP-14 prevented the impairment of HIF-1α transcriptional activity by factor inhibiting HIF-1 (FIH) [66]. In addition, also other transcription factors, such as nuclear factor (NF)-κB, interferon regulatory factor (IRF)-1, and signal transducer and activator of transcription (STAT), play a role in regulating IL-7 transcription. They are activated by type I and II interferons (IFN), IL-12p70, IL-6, IL-1β, and by tumor necrosis factor alpha (TNFα), in a cell-specific and contextual manner [35, 36, 52, 67–71]. Moreover, keratinocyte growth factor (KGF), produced by mesenchymal cells and intraintestinal lymphocytes, stimulates IL-7 expression in colon adenocarcinoma cells by activating the STAT1/IRF-1, IRF-2 signaling pathway [72]. Transforming growth factor (TGF)-β, in turn, has been shown to downregulate IL-7 expression in normal stromal cells [73]. However, it has an opposite effect on pancreatic cancer cells [74]. The upregulation of IL-7 mediated by TGF-β involves epigenetic mechanisms. Specifically, TGF-β activates GLI1, a C2-H2-type zinc finger transcription factor, which forms a complex with SMAD4 and histone acetyltransferase P300/CBP-associated factor (PCAF) [74]. Apart from TGF-β, type I IFNs may also act differently in cancer. They have been shown to downregulate IL-7 expression in tumor bed in murine model of breast cancer [75].

2.3 Interleukin 7 Receptor (IL-7R)

The IL-7 receptor consists of two transmembrane components—CD132 and CD127. Although expressed on the cell surface independently, both subunits are required for successful IL-7 signaling. The CD127 (IL-7Rα) is specific for IL-7, whereas CD132 is a gamma chain (γc), common for IL-2, IL-4, IL-9, IL-15, and IL-21 receptors. The IL-7 receptor is devoid of kinase activity and signals by activating various intracellular kinases upon the cytokine-induced dimerization. The cytoplasmic part of CD127 contains several distinct regions serving as docking domains. The region proximal to cell membrane contains box1 and box2 domains, involved in transferring mitotic and survival signals by binding Janus kinase (JAK)-1. In turn, JAK3 kinase is physically associated with a C-terminal part of CD132. Further away from the membrane, a domain rich in acidic amino acids is located. It facilitates docking of Src kinases, including p53/56^{lyn}, p59^{fyn}, and p56^{lck}. Members of the STAT-family, which are activated downstream of JAKs, bind to the phosphorylated tyrosine residue in the C-terminal receptor region. This region is also necessary for p85/PI3K activation and thus crucial for conveying signals inducing cell proliferation, growth, and survival (reviewed in [12, 76]). The CD127 is present on hematopoietic cells, primarily of the lymphoid lineage, and its expression is strictly regulated to meet the specific needs associated with particular developmental stages (reviewed in [12]). It is abundantly present on precursor and developing immune cells, but only on subsets of mature cells [12, 21, 44, 77]. The responsiveness of malignant lymphocytes to IL-7 is positively correlated with CD127 expression [78]. Most of acute lymphoblastic leukemia (ALL) cells from children with initial and remittent disease express the receptor, although the frequency of CD127$^+$ cells differ by ALL subgroup. It is the highest in pre-B ALL, followed by common ALL, and the least pronounced in pre-T-ALL [76]. Some authors have screened malignant cell lines for IL-7 and/or CD127 expression [4, 71, 79, 80]. However, their findings regarding certain lines have been subsequently undermined [58, 81, 82].

The upregulated expression of CD127 can lead to neoplastic transformation. The AKR/J mice, overexpressing CD127, spontaneously develop thymoma [83]. In addition, combined mutations in CD127 and N-Ras genes have been demon-strated to possess a transforming potential as well [84]. The CD127 promoter sequence contains consensus sites for Ikaros, P.U.1, GATA-binding protein, NF-κB, and Runt-related transcription factor 1 (RUNX1). Moreover, it binds interferon-stimulated (ISRE) and glucocorticoid (GRE) response elements (reviewed in [65]). The IL-7 stimulation triggers rapid CD127 endocytosis via clathrin-coated pits, characterized by decreased receptor recycling and accelerated degradation, involving both lysosome and proteasome-dependent pathways. Importantly, this interleukin-mediated negative regulation of CD127 is crucial for proper activation of JAK/STAT and PI3K/Akt pathways [85]. However, pre-B-ALL cells do not seem to respond to IL-7 stimulation with the receptor downregulation [86].

A number of polymorphisms in CD127 gene have been described (reviewed in [87]) and linked with increased cancer susceptibility [88]. Gain-of-function mutations in CD127, predominantly in exon 6, are found in about 9–12% of pediatric T-ALL [89, 90]. They occur also in adult T-ALL, childhood precursor B cell ALL (B-ALL), and adult acute myeloid leukemia (AML), albeit less frequently [90]. No such observations were made for childhood AML, multiple myeloma (MM), or non-Hodgins lymphomas (NHL) of both T and B-cell origin. In case of solid tumors, CD127 mutations have been found in the colon and non–small cell lung cancers. They were less frequent (<1%) and their nature differed, as they were mostly frameshift and missense mutations. In turn, no mutations have been found in other gastrointestinal tract cancers, in breast and prostate cancers, ovarian stromal tumors, soft tissue sarcoma, or malignant meningioma [90].

Goodwin et al. [91] detected a soluble form of human IL-7Rα (sCD127), capable of IL-7 binding. Subsequently, Crawley et al. [92] showed sCD127 to hamper IL-7/IL-7R signaling, as the activation of two main downstream pathways, JAK1/STAT5 and PI3K/Akt, has been reduced in its presence. A C > T polymorphism (rs6897932; Thr244Ile), favoring a soluble receptor form, has been found to increase susceptibility for breast cancer [88]. In addition, several alternatively spliced CD127 variants encoding truncated but functional receptor have been detected in children with ALL. The truncated receptor is missing domains conveying differential signals, thus promoting mitogenic and survival functionality of IMP II-7/II-7R pathway [70]. Recently, a long noncoding RNA, named lnc-IL7R, has been discovered. It overlaps with the sense strand of the CD127s 3′ untranslated region (3′UTR) [93–96]. The lnc-IL7R presence has been reported in normal peripheral blood mononuclear cells (PBMCs) and malignant monocytic cell line [93]. It has also been detected in oral squamous cell carcinoma [95] and cervical cancer [96] and in fibroblast-like synoviocytes [94]. Noteworthy, despite a close physical relation and some shared functionalities [93–96], the lnc-IL7R has no effect on CD127 expression and seems to function independently [93].

2.4 IL-7/IL-7R Interaction with Extracellular Matrix (ECM)

The interaction with ECM partakes in regulating bioavailability and bioactivity of IL-7. The interleukin preferentially binds to glycosoaminoglycans (GAGs) containing heparin and heparan sulfate [97]. Heparin is a carrier molecule, which additionally protects IL-7 from degradation and prevents accidental activation of random targets [98]. Cell surface GAGs on the interleukin-responsive cells cooperate with IL-7R in IL-7 binding. Those on stromal cells act as weak docking sites, facilitating cytokine positioning and compartmentalization [99]. Sequestration of secreted IL-7 is likely to contribute to the relatively low concentrations of cytokine in circulation (reviewed in [13]). Still, serum concentrations of IL-7 are reportedly higher in cancer patients with various malignancies [2, 5, 100–110] and increasing along with tumor progression [102, 105, 107, 111, 112]. Elevated IL-7 is associated with higher risk for lung cancer [113], as well as with increased likelihood of colorectal adenomas [114] and their greater potential for malignancy

[107]. However, the interleukin association with lymph node involvement [107, 115, 116] or patients' survival [105, 117, 118] remains equivocal. Noteworthy, IL-7 concentration is also higher in aqueous humor from patients with retinoblastoma [119].

Changes in the tensional state of the ECM, with its stretching and folding, alternately hide and reveal binding sites for IL-7. It allows for continuing biological effects of the interleukin, even days after cessation of treatment [120]. The ECM-bound IL-7 is more potent than soluble interleukin, and stromal cell surface molecules might participate in IL-7 activity. As an illustration, biglycan and matrix glycoprotein sc1 have been shown to increase interleukin-dependent proliferation of pre-B cells [121]. The GAGs control the formation of pre-pro-B-cell growth-stimulating factor (PPBSF), a heterodimer consisting of IL-7 and a β chain of hepatocyte growth factor (HGF). The PPBSF allows for complimentary signaling via IL-7R and HGF (c-Met) receptors in cells expressing low levels of CD127 [122]. The IL-7 increases the adhesive interactions of resting and activated T cells with ECM in an integrin-dependent manner, more efficiently if bound to ECM components, especially fibronectin [123]. Moreover, fibronectin induces IL-7 expression in chondrocytes [49]. The IL-7 activates integrins on lymphatic endothelial cells [24] and T cells [124]. It also regulates the expression of adhesion molecules and cadherins on various cancer cells [54, 82, 125, 126], modulating cell-cell and cell-ECM interactions. The interleukin relation with ECM is bidirectional. The IL-7 affects ECM composition by negatively regulating the synthesis of collagen type I, fibronectin, and α-smooth muscle actin (α-SMA) by fibroblasts [127, 128]. In addition, IL-7 facilitates ECM degradation by upregulating MMP-13 [49, 129]. The ECM degradation and remodeling is a crucial step in cancer invasion and metastasis. The interleukin, apart from an indirect MMP-13 upregulation, induces the expression of several other MMPs in various cancer cells [47, 54, 130].

2.5 Overview of IL-7/IL-7R Signaling

Our knowledge on IL-7/IL-7R signaling, even in the most extensively studied lymphocytes, remains fragmentary and the downstream pathways and their molecular end-points are still emerging. Moreover, the pathways are largely redundant and their contribution in particular molecular effect depends on cell type, developmental stage, and context. The interleukin-activated pathways in solid tumors are mostly unknown. The JAK/STAT and PI3K/Akt are the two main signal transduction pathways activated by IL-7/IL-7R and implicated in the key cytokine functionalities. Some authors have also shown the activation of MEK/ERK pathway in response to IL-7/IL-7R signaling, but its significance is poorly understood. In addition, IL-7 has been shown to cooperate with other kinases, adaptor proteins, and transcription factors, for which upstream and/or downstream signaling elements remain to be determined. A simplified overview of IL-7/IL-7R axis is presented in Fig. 2.1. The detailed summary on known IL-7 targets and involved pathways is presented in Table 2.1 (dedicated to hematological cells), Table 2.2 (dedicated to non-hematological cells), and Table 2.3 (dedicated to antitumor immunomodulatory activity of IL-7).

2.5.1 JAK/STAT

The cross-phosphorylation of JAK1 and JAK3 is the first and mandatory step in IL-7/IL-7R signaling. However, some cancer-related mutations facilitate receptor homodimerization and pathway activation independently from γc-associated JAK3 and IL-7 stimulation [89, 131, 132]. They result in constitutive activation of proteins regulating cell cycle progression, survival, and protein translation, as well as in transactivation of other pathways, namely, PI3K/Akt and MAPK [133]. The JAK kinases phosphorylate various STATs, facilitating their dimerization and nuclear translocation, and the type of activated STAT determines the response specificity. The IL-7/

IL-7R signaling via STAT1 has been demonstrated in fibroblasts, in which it inhibited TGFβ synthesis [134] and in B-cell precursor acute lymphoblastic leukemia (BCP-ALL), supporting cell proliferation [135]. In the T-cells, robust STAT1 activation is suppressed by T-cell protein tyrosine phosphatase (PTPN2) [136]. The STAT3-mediated signaling participates in prosurvival activity of IL-7 in B-cell progenitors [137] and in the upregulation of IL-17 secretion by a subset of $\gamma\delta$T cells [138]. However, IL-7 stimulation preferentially induces the JAK-mediated activation of STAT5 [136]. The STAT5 is involved in conveying prosurvival, mitotic, trophic, and differential signals for non-transformed (reviewed in [12, 139]) and malignant lymphocytes T and B [135, 140–144]. However, the pathway relevance for particular functionalities seems to be highly cell-dependent and context-related. As an illustration, IL-7/STAT5 has been claimed to be necessary for T-cell proliferation, but not survival [145]; for differentiation, but not survival or proliferation [146]; or for survival at the expense of proliferation [147, 148]. In addition, STAT5 mediates IL-7 effect on T-cell cytotoxicity [147] and drug resistance [149, 150]. The STAT5-mediated gene regulation employs epigenetic mechanisms, as it increases chromatin accessibility at the promoter regions of target genes and possibly upregulates the expression of histone methyltransferase EZH2 [151–153]. The epigenetic regulation has been implicated in the interleukin-induced Th9 differentiation and IL-9 expression [152], upregulation of Pim1 kinase in T-ALL [153], and in polyfunctionality and stemness in CD4$^+$ T cells [151]. Regarding solid tumors, the STAT5 activation induced by IL-7 has been reported in prostate cancer [154].

The JAK/STAT and PI3K/Akt pathways may be interacting—the STAT5/Pim1 is required to induce and sustain Akt phosphorylation—in order to facilitate cell growth in normal [155] and malignant T cells [140]. The STAT5-mediated Pim1 activation participates also in B-cell development [156]. The Pim serine/threonine kinases are frequently overexpressed in cancer, and their transcriptional activity promotes survival, cell cycle progression, and proliferation and mediates

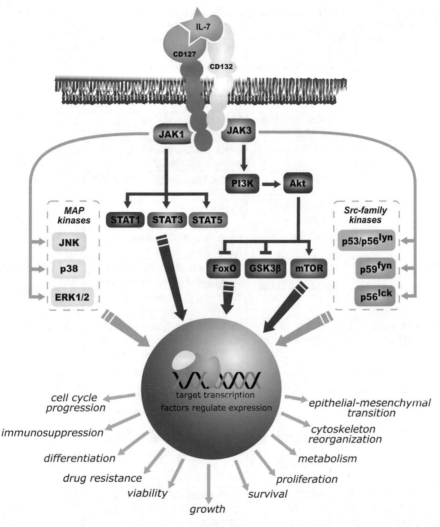

Fig. 2.1 Overview of IL-7/IL-7R signaling cascades. The binding of the IL-7 to the IL-7R elicits phosphorylation events in several signaling pathways. A cross-phosphorylation of JAK1 and JAK3 is the first step. The JAK kinases phosphorylate various STAT proteins, facilitating their dimerization, nuclear translocation, and regulation of gene expression. In another signaling pathway, JAK kinases activate a phosphoinositide 3-kinase (PI3K), the product of which activates the protein kinase B (Akt). Cells stimulated by IL-7 may also activate ERK, JNK, and p38 kinases belonging to the MAPK pathway. Additionally, an IL7-induced activation of p56lck, p53/p56lyn, p59fyn were observed. Various target-transcription factors phosphorylated in signaling pathway activate or repress gene expression of genes leading to many cellular processes. Main pathways induced by IL-7/IL-7R are marked in black, and the other in gray

bone destruction. Among direct Pim1 substrates are transcription factor Myc (activated) and pro-apoptotic Bad (inactivated), as well as the cell cycle regulators: activator CDC25A/C (activated) and inhibitors p21^{CIP1} and p27^{KIP1} (inactivated) (reviewed in [157]). In addition to Pim1, also Pim2 is implicated in IL-7/IL-7R signaling, although its upstream activators have yet to be determined. The Pim2 partakes in IL7-induced osteoclastogenesis in MM by blocking expression of transcriptional targets of BMP2 via antagonizing SMAD1/5 and p38MAPK signaling [158].

The JAK1/STAT5/Pim-1 pathway [140] or a direct phosphorylation by JAK3 [159] activates NFAT2, a transcriptional factor forming com-

Table 2.1 The effects of IL-7/IL-7R signaling on lymphoid and myeloid, normal and malignant cells

Target	Act.	Pathway	Effects of IL-7 stimulation	Cell type	Refs.
↑IL2R	P		↑Proliferation	T/B-ALL, T/B-CLL AML	[209]
	P		↑Proliferation	Lamina propria lymphocytes	[64]
	P	Src kinases	↑p59^{fyn} and p56^{lyn}	hT cells	[183]
	P	RasGRP1-Ras-PI3K/Akt	Mix of IL-7, 2, and 9; relevance-nd	mT-ALL	[176, 214]
↑Pim1; ↑cyclin D2; ↑Bcl-x$_L$	P, S, C	STAT5	↑Proliferation; ↑survival; restored differentiation	Pre and pro-B cells in TGM	[156]
↑IL-2Rα		NFκB/Rel	↓IκBα; ↑p65 and p50 NF-κB/Rel subunits	hPB T cells	[244]
	P	JAK/STAT1; STAT5	↑proliferation	hBCP-ALL; mBAF3 cells	[135]
	P, S	mTORC/p70S6	↑Proliferation; ↑survival; ↑ph-p70 S6	preB-ALL	[211]
↑Bcl-2; ↑Bax	S		↓Apoptosis; ↑Bax (low and uncorrelated with apoptosis rate)	hT-ALL (primary)	[196]
↑Bcl-2	S		↑Survival; (activation, cytotoxicity, IFNγ-na)	hCD56^{bright} NK cells	[77]
↑Bcl-2; ↑cyclin A and D2; ↑cdk4 and cdk2; ↓p27^{KIP1}	C, S	Abolished by rapamycin	↑Viability; hyperphosphorylation of Rb; cell cycle progression to the S and G2/M phase; (Bcl-x$_L$, Bax, Bad-na)	hT-ALL (primary)	[197]
↑Bcl-2; ↑ph-Rb; ↓p27^{KIP1}; ↑GLUT-1	C, S, M$_E$	PI3K/Akt	↑Viability; cell cycle progression; ↑glu uptake, maintained Δψ$_m$; ↑MEK/ERK (relevance-nd)	hT-ALL (primary); TAIL7	[180]
↓p27^{KIP1}	C	Skp2-dep.; PKCθ-dep. (dominant)	Posttrans. reg. via degradation-inducing phosphorylation	IL7–dep. thymocyte line D1; peripheral T cells	[208]
↑CK2; ↑Bcl-2; ↑cyclins A and E; ↓p27^{KIP1}; mild ↓cyclin D2; ↑CD71	P, C, S, G	Jak/STAT; PI3K/Akt	Mediated by CK2: ↑survival; ↑proliferation; maintained Δψ$_m$; ↑cell size	T-ALL (primary); TAIL7, HBP-ALL	[142]
↑ph-S6RP; ↑ph-p53; ↓p27^{KIP1}	P, S, C, G	PI3K/Akt/mTOR	↑Proliferation (DNA synthesis); ↓apoptosis; ↑cell size; ↓% of G1 cells; (Cdk6 and Cdk2, cyclin D3-na)	hT-ALL (primary); TAIL7	[172]
↑Bcl-2; ↓p27^{KIP1}; ↑Ki-67	P, S, V		↑Expansion; ↑leukemia-related death; ↑proliferation; ↑survival; ↑viability	TGM with hT-ALL	[198]
↑ROS	V	Positive loop btw ROS and PI3K/Akt/mTOR	↑Intracellular ROS; maintained Δψ$_m$	hT-ALL (primary); TAIL7	[217]
↓p27^{KIP1}; ↑CD71; ↑Bcl2, ↑Bcl-X$_L$, ↑Mcl-1; ↑Pim1, ↑IKZF4, ↑SOCS2, ↑CISH, ↑OSM; ↓BCL6, ↓IL-10	P, C, G, V	STAT5/Pim1 PI3K/Akt	↑Cell growth; cell cycle progression; ↑proliferation; ↓apoptosis; STAT5-mediated functions: cell growth and proliferation	T-ALL (primary); TAIL7, HBP-ALL	[140, 153]

(continued)

Table 2.1   (continued)

Target	Act.	Pathway	Effects of IL-7 stimulation	Cell type	Refs.
↑GLUT1	G, M_E	STAT5/ Pim1 → Akt	↑glu uptake; ↑GLUT1 surface trafficking	Naive and activated hT cells	[155]
	G, M_L		↑Cell size; ↑glycolytic rate	B23 cells (immortalized mB cell progenitors)	[216]
↑Slc1a4, Slc1a5, Slc7a5 and Slc7a6	G, M_E	PI3K/mTOR	↑Cell size; ↑aa transporters (glu transporters: Slc2a1, Slc2a3, Slc2a9 and Slc5a2-na)	Naive mCD8+ T cells	[219]
↑Aquaporin 9	S, M_E		↑Glycerol uptake; ↑TAG synthesis; ↑survival	Memory h/mT cells	[220]
↑Myc; ↑chol&aa metabolism; ↑E2F; ↓neg. reg. cell cycle	G, M_E, P, C_Y	IL-7R/mTOR; JAK/STAT5	↑Anabolism; ↑proliferation; ↑DNA synthesis; ↑cell growth; ↑cell cycle progression; ↓cytoskeletal reorganization	proB cells; TGM (Myc-induced B cell lymphoma)	[143]
↑Crk; ↑STI1; ↑ATIC; ↑hnRNPH; ↑Myc; ↑lymphotoxin α; ↑PKCη; ↑SOCS2; ↓LAIR1; ↓Flt3	S, P, E_M	STAT5	Crk-mediated: ↑cell survival; ↑ heterodimerization of Crk and STAT5	IL7–dep. thymocyte line D1; murine pro-B cells; hT cells	[166]
↑Bcl-2	S, G	PI3K/Akt/ mTOR (atrophy)	↑Survival (via Bcl-2); prevented atrophy, maintaining cell size and metabolic rate (not by Bcl-2); (proliferation-na)	Naive mT cells	[193]
↑Bcl-2; ↑Bcl-x_L; ↑NFATc3; ↑SOCS3; ↑Tcf12, Tcf3, Rag1, Rag2 and Ptcra	S, D	JAK3/NFAT2	Calcineurin-independent activation of NFAT2	Pre-TCR thymocytes	[159]
↑Bcl-2, ↑c-Myb	S	STAT5/	↑ c-Myb DNA binding and expression (c-Myc-na); ↑STAT5 binding to Bcl-2 promotor	CTCL cell line SeAx	[141]
↑Bcl-2; ↑Bcl-x_L; ↓Bax	S	pH stabilization	↑Survival; (Bak, Bad-na); maintains Bax in cytosol	IL7–dep. thymocyte line D1	[194]
↑Bcl-2; ↓Bax	S	(of p53 and Fas/FasL -not involved)	↑Survival; (Bcl-x_L, Bad, Bcl-w; proliferation-na)	Murine pro-T cells	[195]
↑Bcl-2	S	STAT5	↑Survival; (Mcl-1, Bcl-x_L-na)	Murine recent thymic emigrants	[148]
↑IFNγ, FasL	S	STAT5	Continuous IL-7 stimulation leads to CICD	Naive CD8+ T cells	[205]
↑ICOS	S	NFκB	NFκB as a negative regulator of IL-7 signaling	NKT cells	[252]
↑ph-Pyk2	S	JAK1	↑Survival	IL7–dep. thymocyte line D1	[191]
Bad↓	S	PI3K/Akt	Posttrans. reg. via ↑Bad phosphorylation-no mitochondrial translocation	IL7–dep. thymocyte line D1	[200]
↓Bim	S		Posttrans. reg. via ↑Bim phosphorylation-resulting in altered pI of Bim_{EL} isoform	IL7–dep. thymocyte line D1; peripheral T cells	[201]
↑Mcl-1	S		Protein and mRNA expression	Thymocytes or purified T cells	[199]

(continued)

Table 2.1 (continued)

Target	Act.	Pathway	Effects of IL-7 stimulation	Cell type	Refs.
↑CD47	I_S		Immune evasion by ↓phagocytic functions; (proliferation, apoptosis-na)	Sézary cells from PB	[263]
↑PD-1; ↑PD-L1	I_S		↓T-cell effector functions: ↓IL-2 and IFNγ expression; (proliferation, survival-na)	PBMCs, CD4$^+$ and CD8$^+$ T cells, monocytes/macrophages	[257]
↑IL-6	D_R, I_S		↑clonogenic growth; ↓apoptosis; partially protects from doxorubicin cytotoxicity; costimulator for Tregs proliferation	Cultured H-RS cells, HL-derived fibroblasts	[57]
↑AUTS2; ↓MEF2C		STAT5	Synergistic inputs of AUTS2 and MEF2C in lymphopoiesis and leukemia (de)regulating NKL homeobox gene MSX1	T-ALL	[144]
↑ph-YB-1	D_R		↑Survival in rapamycin-treated cells	BCP-ALL	[181]

Act activity, *Ref* reference, *h (prefix)* human, *m (prefix)* murine, *C* cell cycle progression, C_Y cytoskeleton reorganization, *D* differentiation, D_R drug resistance, E_M epithelial-mesenchymal transition, *G* cell growth, I_S immunosuppression, M_E metabolism, *P* proliferation, *S* survival, *V* viability, *T-ALL* T-cell acute lymphoblastic leukemia, *B-ALL* B-cell acute lymphoblastic leukemia, *T-CLL* T-cell chronic lymphoblastic leukemia, *B-CLL* B-cell chronic lymphoblastic leukemia, *AML* acute myeloid leukemia, *BCP-ALL* B-cell precursor acute lymphoblastic leukemia, *CTCL* cutaneous T-cell lymphoma, *H-RS cells* Hodgkin and Reed/Sternberg cells, *HL* Hodgkin lymphoma, *TGM* transgenic mice, *na* not affected, *nd* not determined, *PB* peripheral blood, *PBMCs* peripheral blood mononuclear cells, *dep.* dependent, *ph (prefix)* phosphorylation, *posttrans.reg.* posttranslational regulation, *CK2* casein kinase 2, $\Delta\psi_m$ mitochondrial membrane potential, *glu* glucose, *aa* amino acids, *TAG* triacylglycerols, *CICD* cytokine-induced cell death, *neg. reg.* negative regulators, *btw* between

plexes with AP-1, GATA, Foxp3, or MEF, in a noncanonical, calcineurin-independent route. In addition, IL7-mediated activation of Akt, and subsequent GSK3 inhibition, might upregulate NFAT2 activity by preventing its translocation from the nucleus [160]. The potential NFAT2 targets include a wide panel of cytokines and growth factors with proinflammatory and angiogenic activity, prosurvival and pro-proliferative Mdm, cyclin D1, and Bcl-2, and transcription factors involved in metabolism regulation, such as c-Myc, HIF-α, and IRF4 [160]. Up-to-date, a stimulatory JAK3/NFAT2-mediated effect of IL-7/IL-7R on cell survival (via Bcl-2 and Bcl-x$_L$) and differentiation (via Tcf12, Tcf3, Rag1, Rag2, and Ptcra), as well as on NFAT4 and SOCS3 has been demonstrated [159]. The NFAT2 may also mediate IL7-induced expression of PD-1 on CD4$^+$ and CD8$^+$ T cells [161]. As NFAT2 is commonly expressed in lymphoid cells, its aberrant activation in lymphomas/leukemias is to

be expected. However, NFAT2 overexpression is associated also with pancreatic and colorectal cancer and linked with tumor invasion and metastasis [160, 162]. The proposed mechanism of acquiring the ability to express NFAT2 is a fusion between tumor cells and NFAT2-expressing cells from tumor microenvironment, such as lymphocytes and myeloid cells, and trading of genetic material [163]. Indeed, the formation of such hybrid cells has been confirmed in vitro, as well as in vivo [164].

Apart from the canonical JAK/STAT activation, early studies have pointed at possible JAK3-mediated phosphorylation of p85 and activation of PI3K/Akt [165]. The notion has recently been corroborated by constitutive activation of the PI3K/Akt pathway in T-ALL cells harboring mutations in JAK3 gene [133].

The JAK/STAT pathway is inhibited by a family of SOCS proteins. They are activated by IL-7/IL-7R signaling, constituting a classic negative

Table 2.2 The effects of IL-7/IL-7R signaling on non-hematological cells and on osteoblastogenesis and osteoclastogenesis

Target	Act.	Pathway	Effects of IL-7 stimulation	Cell type	Refs.
–	P	PI3K; JAK3	↑Proliferation	MDA MB231, MCF-7 (BC)	[46]
↑Cyclin D1	I, C	AP-1 (c-Fos/c-Jun)	↑Proliferation, G1/S transition, ↑tumor growth; (cyclin C and E-na)	MB??, LH7 (LC), XT	[310]
↑Cyclin D1; ↓p27^{kip1}	P, C	PI3K/Akt	↑Proliferation; ↓% cells in G1 and ↑% cells in G2/M phase	MDA MB231, MCF-7 (BC)	[174]
↓p21^{Cip1}	C, I	–	IL7R KO-induced G2 cell cycle arrest, ↑p21, ↓wound healing capacity	HCE7, HCE4 (ESCC)	[9]
↑MMP-9, ↑cyclin D1, ↑E-cadherin; ↓vimentin, ↓βcatenin	C, I, E$_M$	Akt; JNK	↑Migration; ↑proliferation; ↓EMT; (STAT5-na)	HPV-transf. hepatoma	[54]
↑ZEB1/2, ↑TWIST1, ↑SNAI, ↑N-cadherin, ↑vimentin; ↓E-cadherin	I, E$_M$	STAT5; PI3K/Akt; ERK	↑Wound-healing migration; ↑invasion; ↑tumor sphere formation; ↑EMT markers; metastases less responsive; ↑IL7R promotes bone metastasis; (proliferation-na; MMPs-na)	PC-3 (PC); bone/liver metastases in XT	[154]
↑MMP-9, ↑MMP-2; ↑p27^{KIP1}	I, C	p21^{KIP1}/ERK1-2/NFκB; AP1	↑Wound healing migration; ↑invasion; ↑DNA binding by NFκB and AP1; (SP1-na; proliferation-na; other cell cycle proteins-na)	5637 (BlC)	[130]
↑MMP-3, ↑MMP-7	I, M	Akt/NF-κB	↑Invasion; ↑migration; (viability-na)	LNCap, PC-3, and DU-145 (PC); RPWE-1 (N)	[47]
↓COL1A1, ↓COL3A1	E$_C$	–	↓ECM synthesis; ↑Smad7; ↓ph-PKCδ; ↓TGFβ signaling	hPF	[127]
↓TGFβ, ↓COL1A1, ↓FN, ↓αSMA	E$_C$, M	STAT	↓Migration; ↓ECM synthesis; ↑Smad7; ↓ph-PKCδ; ↓TGFβ signaling	hSF	[128]
↑MMP-13	E$_C$	Pyk2	↑ECM degradation	hCh	[49]
↑MMP-13; ↑S100A4	E$_C$	JAK/STAT3	↑ECM degradation; ↑S100A4 → ↑MMP13 (via RAGE)	hCh	[129]
↑Bcl-2; ↓Bax, ↓p53	S	p53	↓Apoptosis in HBE and A549; (H1299-na, lack of p53)	A549, H1299 (LC); HBE (N)	[202]
↓E-cadherin; ↑N-cadherin	E$_M$	PI3K/Akt	↑Scattered morphology; ↑invasion; ↑EMT; ↑lung meta.; ↓overall survival	MCF-7, BT-20 (BC); GT	[82]
↑VEGF-A	A$_G$	–		mTC-EC	[228]
↑podoplanin, ↑prox-1, ↑LYVE-1, ↑VEGF-D, ↑VEGFR3	L	PI3K	↑Cell growth; ↑migration; ↑microtubule in vitro; ↑lymphatic tubule in vivo (vWF, VEGF-A, VEGFR1/2-na; little ↑VEGF-C; JAK/STAT-na)	HECV; mice with HECV±IL-7	[175]
↑VEGF-D	L	–	↑Microtubule formation in vitro; ↑formation of lymphatic LYVE-1+, but not vascular vW+, microtubules in vivo (VEGF-A/B/C-na; not in BT-483)	MDA MB231, MCF-7, BT-483 (BC); GT/MDA MB231 ± IL-7	[230]
↑VEGF-D	L	AP-1 (c-Fos/c-Jun)	↑ex: c-fos and c-jun; ↑ph: c-jun; ↑dimerization; ↑AP-1 DNA binding; ↑lymphatic tubules; (migration-na; VEGF-A/C-na)	A549, SPC-A1, LH7, SK-MES-1 (LC); GT	[6, 81]

(continued)

Table 2.2 (continued)

Target	Act.	Pathway	Effects of IL-7 stimulation	Cell type	Refs.
↓RUNX2/CBFA1	O_B	–	↓RUNX2/CBFA1 DNA; ↓osteoblastogenesis	MMCs (IL-7)—BMSC and preOBs cell line (RUNX2/CBFA1)	[278]
↑Pim-2	O_B	–	Pim-2 → ↓BMP-2; ↓osteoblastogenesis	BMSC and pre-OBs	[158]
↑Gfi1	O_B	–	↑Gfi1 protein; ↑nuclear location→↓Runx2; ↓OBs differentiation	MMCs	[190]
↑RANKL, ↑M-CSF	O_C	–	↑Osteoclasts formation	hPBSCs	[35]
↓OPG	O_C	–	↓Transcriptional activity of Runx2/Cbfa1	ROS 17/2.8 (rOS)	[277]
↑RANKL	O_C	–	↑Osteoclastogenesis	MMCs (IL-7)—T cells (RANKL)	[36]
↑TNFα	O_C	–	↑Osteoclastogenesis	PBMCs from cancer patients	[112]
↑IL-17A	I_S, T	–	↑Tumor growth; ↑lung meta.; ↑IL-17A in tumor; ↑tumor-infiltrating γδT cells	Murine BC	[75]
↑IL-17	I_S, T	STAT3	↑Tumor growth; ↑MDSCs in tumors; ↑IL17A⁺ γδT cells	GT/B16-F10 (Mel)	[138]
–	D_R	–	↑Cell growth; ↓apoptosis	CP-gliomas	[4]
↑Beclin-1	A	–	Beclin-1 suppression ↑PI3K/Akt/mTOR	A549 (LC)	[58]
↑ICAM	–	–	–	A375 (Mel); G361 (N)	[126]

Act activity, *Ref.* reference, *A* autophagy, A_G angiogenesis, *C* cell cycle progression, D_R drug resistance, E_C extracellular matrix, E_M epithelial-mesenchymal transition, *I* invasion, I_S immunosuppression, *L* lymphangiogenesis, *M* migration, O_B osteoblastogenesis, O_C osteoclastogenesis, *P* proliferation, *S* survival, *na* not-affected, *KO* knockout, *BC* breast cancer, *GT* tumors in mice from grafted cancer cells, *ESCC* esophageal squamous cell carcinoma, *PC* prostate cancer, *BlC* bladder cancer, *mTC-EC* murine thymic cortical epithelial cell line, *HECV* human endothelial cell line displaying characteristics of vascular and lymphatic endothelium, *N* normal, *transf.* transfected, *meta.* metastases, *hPF* human pulmonary fibroblasts, *hSF* human subconjunctival fibroblasts, *BMSC* bone marrow stromal cells, *pre-OBs* preosteoblasts, *MMCs* multiple myeloma cells, *PBMCs* peripheral blood mononuclear cells, *hPBSCs* peripheral blood stem cells, *rOS* rat osteosarcoma, *Mel* melanoma, *CP* cisplatin-treated, *ex* expression, *ph* phosphorylation, *FN* fibronectin, *αSMA* α-smooth muscle actin, *hCh* human chondrocytes

feedback loop [140, 159, 166–168]. The SOCS proteins induce ubiquitination and proteasomal degradation of STATs. In addition, they might directly inhibit JAKs [169] and/or induce ubiquitination and proteasomal degradation of CD127 [167]. The SOCS proteins are frequently mutated and silenced in cancer, especially in hematological malignancies [169].

2.5.2 PI3K/Akt

Upon IL-7/IL-7R stimulation of JAK1/3, the p85 subunit of PI3K is phosphorylated, activating Akt kinase. The main downstream Akt targets relevant for IL7-mediated signaling are GSK3 (inhibited), FoxO (inhibited), and mTORC (activated). The interleukin-induced activation of PI3K/Akt promotes cell survival, proliferation, and growth by inhibiting Bad, Bim, Bax, p21^{CIP1}, and p27^{KIP1}, and activating Cdk2. The members of PI3K/Akt pathway, as well as its inhibitor—phosphatase and tensin homolog (PTEN), are frequently mutated in cancer, resulting in pathway hyperactivity. The strongest level of activation is conferred by JAK1 mutations [170].

Malignant cells in T-ALL relay on PI3K/Akt for antiapoptotic, mitotic, and trophic stimuli. The rates of cell viability, survival, and proliferation are largely reduced by rapamycin, implying mTORC involvement [142, 170–172]. The pathway seems to be more important for the development of lymphocytes T than B. In fact, global analysis of events triggered by IL-7/IL-7R signaling in pro-B cells has shown mTORC1/Myc activation without PI3K/Akt involvement.

Table 2.3 Antitumor activities of IL-7/IL-7R signaling

Mediators	Effect of IL-7 stimulation	Cell type	Refs.
↑IL2R	↑Proliferation; phenotypical changes: ↓CD56⁺, ↑CD3⁺ with CD4⁺ dominance	TILs from renal cell carcinoma	[243]
	↑CD8⁺ Tc: tumor rejection	IL7⁺ murine glioma cells grafted to mice	[234]
	↓Tumorigenicity; tumor rejection and resistance to re-challenge or retarded growth; ↑CD4⁺ and CD8⁺ Tc infiltration; ↑cytotoxicity; ↑basophils, eosinophils	IL7⁺ murine fibrosarcoma grafted to mice	[237]
↑IFNγ	↑TILs growth and toxicity; ↑IL2-mediated LAK activity	TILs from renal cell carcinoma	[240]
↓TGFβ	Retarded tumor growth for M-24; ↑lymphocyte cytotoxicity; (proliferation-na)	Human melanoma cell lines M-14 or M-24 (IL7⁺) grafted to mice	[56]
↓TGFβ	Transcriptional regulation; ↓mRNA and secreted protein	Murine macrophages	[253]
↓TGFβ	↓Immunosuppression: ↓inhibition of lymphocyte proliferation	IL7⁺ murine fibrosarcoma; splenic lymphocytes	[226]
	↑Sensitivity to effector cells; (proliferation, cytokine secretome profile, ICAM, HLA class I and II—na)	Primary malignant melanoma (IL7⁺)	[227]
↑IFN-γ, IL-12, CXCL9, CXCL10;↓PGE2, VEGF-A, TGFβ (t,s)	DC + IL-7 potentiated IL-7 effects: ↑Th1 and antiangiogenic responses; ↓immunosuppressive mediators; ↓tumor burden; ↑survival	Spontaneous murine bronchoalveolar cell carcinoma (lung cancer model)	[231]
↑FasL	↑effector mechanisms; (TRAIL-na)	Primary human NK cells	[206]
↑ICAM; ↓TGFβ	↑Sensitivity to LAK cells, (proliferation, cell cycling, or apoptosis rate-na); ↓tumor growth and dissemination	Human ovarian carcinoma cell line, SKOV3 IL7⁺; xenograft tumors	[125]
↓Bim, ↓Foxp3, ↑SMAD7 (Tc); ↑IFNγ, IL12, MIG, IP10 (t,s,ln);↓TGFβ and IL-10 (t,s)	↓tumor burden; ↑CD4⁺ and CD8⁺ Tc, ↓their apoptotic rate and ↑cytolytic act.; ↓inhibitory act. of Tregs	Mice bearing established Lewis lung cancer	[203]
↑MIG, IP10, IFNγ, IL12; ↓IL10 and TGFβ (t,s); ↑IL12, iNOS but ↓IL10 and arginase (m)	IL-7/IL-7Rα-Fc chimeric molecule: ↓tumor burden; ↑TAM-M1, NK, Tc frequency; ↑cytotoxicity	Mice bearing established Lewis lung cancer	[248]
	↑Tumor-free survival, ↑CD4⁺ and CD8⁺ Tc, ↑CD19⁺ Bc, ↑necrotic areas in the tumor	Immune-competent mouse prostate cancer model; vaccination with IL7-producing whole cells	[251]
	↓Tumor size	Mice inoculated with N32 glioma cells and grafted with MSCs transfected with IL-7, with or without additional IFNg peripheral immunotherapy	[238]
	IL-7 + OXP: ↓tumor cell proliferation; ↑tumor cell apoptosis; ↓metastatic nodules; ↑tumor-infiltrating activated CD8⁺ Tc; ↓splenic Tregs; (TAMs, TADCs, MDSCs—na)	Lung and abdomen metastasis models of murine colon cancer	[232]

(continued)

Table 2.3 (continued)

Mediators	Effect of IL-7 stimulation	Cell type	Refs.
↑SMAD; ↓Foxp3;↑sIL6, sIFNγ; ↓sIL10, ↓sTGFβ	↓Tumor growth; ↑survival; ↑CD4⁺ and CD8⁺Tc; ↑cytotoxicity; ↑Th1/Tc1-type IFNγ-producing effector Tc; ↓Tregs (th,s,ln)	Meth A fibrosarcoma grafted to mice	[242]
↑IFN-γ	↓Tumor growth, ↑IFN-γ and ↑breast cancer cells-specific CTL cytotoxicity	Xenograft model of breast cancer	[241]
(1) ↑IP10, CCL3, CCL4, CCL5, IL-1β, IL-6, TNFα; (2) V-CAM	↑Infiltration of cervicovaginal tissue with CD4⁺ and CD8⁺ Tc, γδTc, DCs in response to IL-7fused with Fc; tumor suppression	(1) cervicovaginal epithelial cells; (2) vascular endothelial cells; orthotopic cervical cancer model	[249]
↑IFNγ, IL-2, TNFα, granzyme B	↑Proliferation and accumulation; ↑polyfunctionality and stemness in CD4⁺Tc; ↑EZH2⁺CD4⁺Tc; ↑H3 acetylation; ↑cytotoxicity; involves STAT5 and epigenetic regulation	Antigen-stimulated murine CD4⁺ Tc	[151]
↑IL-9 and IL-21	↓Susceptibility to develop lung melanoma in vivo by ↑IL-9 expression (t); ↑CD4⁺ Tc differentiation into Th9 with ↑antitumor activity; ↑histone acetyltransferases; engaged PI3K/Akt/mTOR and STAT5	Tumor-specific CD4⁺ T helper 9 (TH9) cells; B16-OVA melanoma cell lines grafted to mice (lung melanoma model)	[152]

Ref reference, *Tc* T-cells, *Bc* B-cells, *TILs* tumor infiltrating lymphocytes, *na* not affected, *t* in the tumor bed, *s* in the spleen, *ln* in the lymph nodes, *m* macrophages, *th* thymus, *OXP* oxaliplatin, *TAMs* tumor-associated macrophages, *TADCs* tumor-associated dendritic cells, *MDSCs* myeloid-derived suppressor cells, *CTL* cytotoxic lymphocytes T, *TAM-M1* tumor-associated macrophages of M1 phenotype, *act.* activity, *s (prefix)* secreted

Moreover, pathway activation was capable of inducing lymphoma in experimental animals [143]. Furthermore, PI3K/Akt suppression by PTEN is necessary for pro-B cell development. It rescues FoxO1-regulated CD127 expression and is required for the interleukin-mediated STAT5 activation and expression of STAT5-regulated genes [143]. This novel IL-7/IL-7R signaling pathway is involved in cell proliferation and growth and in metabolism regulation, but dispensable for survival. Its inhibition protects experimental animals from Myc-induced lymphoma [143].

The PI3K/Akt pathway is also implicated in drug resistance in T-ALL [173]. In solid tumors, pathway involvement in proliferation [46, 54, 174], migration and invasion [47, 54, 82, 154], lymphangiogenesis [175], autophagy [58], and epithelial-mesenchymal transition (EMT) [54, 82, 154] has been reported.

Stimulated by a mixture of IL-7, IL-2, and IL-9, the PI3K/Akt pathway may be activated by RasGRP1, a Ras activator, which typically induces Ras-Raf-MEK-ERK in response to TCR stimulation. The pathway usually involves PLCγ1 activation and an increase in the concentration of intracellular calcium. However, basal levels of diacylglycerol are sufficient for Ras activation following IL-7 stimulation and PLCγ1 is not involved in this noncanonical RasGRP1-Ras-PI3K/Akt axis [176].

2.5.3 MAPK Pathway

Early studies in mice have shown that IL-7 in T cells activates the stress-activated protein kinase (SAPK)/c-Jun N-terminal kinase (JNK) and p38MAP kinases. The inhibition of MAPK-activating protein kinase-2, their downstream effector, has abrogated the interleukin-induced proliferation, but had no effect on c-Myc expression [177]. In turn, it has not affected Shc and p42$^{MAP/Erk}$ kinase, implying that the MEK/ERK pathway is not utilized by IL-7/IL-7R axis [178]. However, human primary CD8⁺ T cells [147] or B cells [156] do not respond to IL-7 with activation of any of MAPK pathways, although Fleming et al. [179] showed that ERK kinases are activated by concurrent IL-7R and pre-B cell receptor (pBCR) stimulation. The mechanism is necessary for pBCR-mediated proliferation of pro-B cells. At low or terminal conditions. Unlike in human primary T cells, IL-7/IL-7R signaling activates MEK/ERK in malignant lymphocytes. The MEK/ERK pathway may promote cell cycle progression and prevents apoptosis by regulating the expression of c-Myc, cyclin D1, $p27^{KIP1}$, and $p21^{CIP1}$. However, its relevance in T-ALL remains uncertain, as the pathway inhibition has no effect on cell viability or cell cycle progression [180]. Instead, the MEK/ERK is implicated in steroid resistance of T-ALLs, via downregulation of proapoptotic Bim [173]. Correspondingly, IL-7 rescues BCP-ALLs from rapamycin-induced apoptosis, acting by upregulating MEK/ERK/RSK2. Consequently, it leads to phosphorylation of YB-1, a marker of multidrug resistance [181]. Moreover, Martelli et al. [182] suggested a role of the interleukin-activated MER/ERK signaling in inducing eIF4B and protein translation. Regarding solid tumors, the IL-7/IL-7R axis activates JNK in hepatoma cells [54] and ERK in bladder [130] and prostate [154] cancer cells, possibly contributing to their migratory properties. The interleukin-induced ERK phosphorylation could be observed in murine model of lung melanoma as well, but its relevance has yet to be determined [152].

2.5.4   Other

The IL-7 activates Src kinases $p56^{lck}$ and $p59^{fyn}$ in unstimulated and stimulated human T cells, contributing to their basal proliferation [183]. In cancer, the $p56^{lck}$ is overexpressed in murine [184] and some human [185] T-ALLs. Moreover, the animals constitutively expressing Src reproducibly develop thymic tumors [184]. The interleukin induces $p59^{fyn}$ and $p53/p56^{lyn}$ activation in pre-B [186] and MM cells [187, 188]. In MM, the $p53/p56^{lyn}$ upregulates expression of MUC1 protein and respective mRNA. The Lyn-mediated phosphorylation of MUC1 on Tyr-46 enhances MUC1 binding to β-catenin and facilitates nuclear translocation of the complex [187]. The MUC1 may also bind to NF-κB p65 and translocate to the nucleus or associate with heat shock proteins and translocate to mitochondria. The effect of its activity is cell renewal and survival [189]. The cross talk between IL-7/IL-7R axis and Wnt/β-catenin pathway may contribute to cancer-promoting functionalities of IL-7, including mitogenic activity (via Jun, c-Myc, and cyclinD-1) and inducing metabolic switch from oxidative phosphorylation to lactate-generating glycolysis. It may also mediate the interleukin's effect on beclin-1 and autophagy, as β-catenin has been shown to inhibit both [190].

IL-7/IL-7R has been demonstrated to induce Pyk2 phosphorylation in a JAK1-dependent manner in normal T cells and is required for their interleukin-mediated survival [191]. In chondrocytes, IL-7 is activating Pyk2, evoking catabolic processes in ECM [49].

2.6   Impact of IL-7/IL-7R Signaling on Cancer Development

2.6.1   IL-7/IL-7R Axis in Cell Survival

The inhibition of apoptosis seems to be the most critical role of IL-7/IL-7R signaling in cells from T-cell lineage, while for B lineage, it is rather proliferation and differentiation [192]. Survival-promoting activity of IL-7/IL-7R axis is mainly associated with the interleukin ability to modulate the expression, phosphorylation status, and location of the Bcl-2 family members—antiapoptotic Bcl-2, Bcl-x_L, and Mcl-1 and proapoptotic Bad, Bim, and Bax. The upregulation of Bcl-2 has been consistently reported in nonmalignant [148, 159, 193–195] and malignant T cells [140–142, 180, 196–198] and in mature NK cells [77]. In turn, the induction of Bcl-x_L has been observed in normal pro-B cells [156], normal developing T cells [159, 194], and in T-ALL [140], although not unanimously [148, 195, 197]. Likewise, Mcl-1 has been seen either upregulated in T-ALL [140] and normal T cells [199] or not affected [140, 148]. The IL-7/IL-7R axis does not seem to have an effect on Bak [194] and Bad [194, 195, 197], although Li et al. [200] reported Bad's phosphorylation and retention in the cytoplasm in response to IL-7. The interleukin downregulates Bim, claimed to be a major inducer of apoptosis in mature T cells [192]. The possible mechanism involves posttranslational modification through subtle changes in protein isoelectric point [201]. Regarding Bax, contrary to T-ALL [197], the protein transcription is downregulated by IL-7 in nonmalignant T cells [195]. In addition, the IL-7/IL-7R axis might induce Bax inhibition through maintaining the right pH. It would prevent the membrane-seeking domains of Bax from being exposed and, consequently, thwarting its translocation to mitochondria [194].

In solid tumors, stimulatory effect of IL-7 on Bcl-2 and inhibitory on Bax and p53 have been demonstrated in lung cancer [202]. In turn, the downregulation of Bim in tumor-infiltrating lymphocytes contributes to antitumor immunomodulating activities of IL-7 [203]. The interleukin-mediated inhibition of p53 phosphorylation has also been seen in T-ALL [172]. The NKT cells are protected from apoptosis by the interleukin-induced ICOS expression. However, contrary to its declared prosurvival activity, IL-7 has also enhanced steroids-induced death in one T-ALL line [204]. Moreover, Karawajew et al. [196] observed a slow increase in proapoptotic Bax upon IL-7 stimulation in primary pediatric T-ALLs. Those observations may suggest that the relationship of the IL-7/IL-7R axis with apoptosis is not as simple as previously believed. Accordingly, constitutive activation of IL-7/IL-7R causes death of nonmalignant T cells, associated with increased FasL synthesis mediated by STAT5. An inhibitory TCR signaling is required for the stimuli to be intermittent and advantageous [205]. Moreover, the upregulation of FasL by IL-7 enhances cytolytic activity of human NK cells [206].

Both PI3K/Akt [140, 148, 180, 197] and JAK/STAT5 [141, 147, 148] are implicated in conveying survival signals from IL-7/IL-7R. Their relevance strongly depends on cell type, its developmental stage, and microenvironmental context. As an illustration, the PI3K/Akt activa-

tion is critical for the interleukin-mediated survival of T-ALL cells [140, 180, 197]. Still, the JAK/STAT5 is involved in protecting T-ALLs against death induced by steroids or histone deacetylase (HDAC) inhibitors [150]. In addition, direct JAK3 phosphorylation of NFAT2 has been shown to induce expression of antiapoptotic Bcl-2 and Bcl-x_L [159]. The involvement of other mediators, downstream main pathways, has been documented as well, including Pim1 [156] and JunD activation of c-Myb in CTCL [141, 207].

2.6.2 IL-7/IL-7R Signaling in Cell Proliferation and Cell Cycle Progression

The mitogenic function of IL-7 is separate from its antiapoptotic activity. Accordingly, it has been demonstrated that Bcl-2 may replace the interleukin during IL-7 withdrawal in providing survival, but not pro-proliferative, stimuli [208]. The interleukin reportedly stimulates proliferation in acute and chronic lymphoblastic and myeloid leukemias [209] and in the lung, breast, and liver cancers [46, 54, 174, 210], but not in prostate and bladder carcinomas [130, 154]. Unlike in T cells, the ensuring survival of solid tumor cells requires high interleukin concentrations [147]. The IL-7/IL-7R increases proliferation by facilitating cell cycle progression through G1/S and G2/M checkpoints [197]. It is achieved mainly by inhibiting cyclin kinase inhibitor p27^{KIP1}, downregulated by IL-7 in T-ALL [140, 142, 172, 180, 197, 198] and in the breast [174] and esophageal cancers [9]. In hepatoma, IL-7 has been shown to downregulate also p21^{CIP1} [54]. The interleukin-mediated regulation of p27^{KIP1} involves its decreased phosphorylation by two distinct mechanisms: reduction of Skp2 and Cks1 protein levels and inhibition of PKCθ [208]. Interestingly, IL-7 has been shown to upregulate p27^{KIP1} in bladder cancer. However, p27^{KIP1} in this setting serves as an enhancer of NF-κB DNA binding, facilitating MMP-9 expression and cancer cell migration [130]. The IL-7 has been observed to upregulate cyclin D2 in nonmalignant and malignant lymphocytes [156, 197, 210] and cyclin D1 in solid tumors [54, 174,

210]. Noteworthy, a mild downregulation of cyclin D1 in T-ALL has been noted as well [142]. In addition, IL-7 induces cyclin A [142, 197] and E [142] and cyclin-dependent kinases cdk2 and cdk4. However, the interleukin affects more markedly their activity than expression [197]. The IL-7 leads to the hyperphosphorylation of retinoblastoma protein (Rb) and release of E2F transcription factors [197]. Only recently, the IL-7/IL-7R signaling has been shown to control over 800 genes, many of which are associated with proliferation, DNA synthesis, and RNA processing [143]. In conveying mitogenic signals, both the JAK/STAT [135, 140, 142, 143, 166] and PI3K/Akt pathways [46, 54, 142, 172, 174, 180] are involved. The mediation of Src [183], Pim-1 [140], mTORC1 [143, 197, 211], Crk-induced LARK1 receptor [166], JNK [54], AP1 [6, 130, 210] mel-18 [212], bmi-1 [213], Skp2 and PKCθ [208], and of non-canonical RasGRP1-Ras-PI3K/Akt pathway [176, 214] has been implicated as well.

2.6.3 IL-7/IL-7R Signaling in Cell Growth and Metabolism

The IL-7/IL-7R axis prevents death in neglected T cells by preserving the rates of glycolysis and respiration and not by regulating Bcl-2 family members. This trophic activity of IL-7 is separate from its antiapoptotic function and is fully dependent on PI3K/Akt/mTOR [193]. The common characteristic feature of cancer cells and activated lymphocytes is their accelerated uptake of glucose and the switch from oxidative phosphorylation to lactate-generating glycolysis, referred to as the Warburg effect. Among many proposed advantages of the phenomenon for cancer cells is the acidification of tumor environment, which enables their invasion and metastasis by facilitating ECM degradation [215]. The IL-7 has been shown to be a key regulator of glucose uptake by T cells. It enhances GLUT1 membrane trafficking in a STAT5-dependent manner, followed by Pim and Act activation. Although delayed, Act induction and sustained activation are critical for IL-7 trophic role [155]. The interleukin controls

cell size and glycolytic rate in B-cell progenitors as well [216]. In malignant T-ALL cells, IL-7 additionally induces GLUT1 expression and increases mitochondrial potential and integrity. Both activities are dependent on PI3K/Akt pathway [180]. Moreover, the IL-7/IL-7R signaling increases intracellular reactive oxygen species (ROS) in T-ALL, which is required to maintain mitochondrial membrane potential and thus is critical for the cell viability. The interleukin upregulates ROS generated by both NADPH oxidase complex (NOX) and electron transport chain (ETC). However, only ETC-generated ROS are necessary for activation of PI3K/Akt/mTOR pathway. Upon activation, the pathway sustains ROS generation by ensuring constant uptake of glucose and by upregulating GLUT1 expression and cell membrane trafficking [217]. The potential effect of IL-7 on GLUT1 and glucose uptake in solid tumors has not been tested, although carcinomas also upregulate the transporter expression. The GLUT1 is mostly undetectable in normal epithelial cells, but its expression correlates positively with cancer dedifferentiation, invasion and metastasis, and, consequently, with poor prognosis [215].

The activation of Akt inhibits FoxO1, a negative regulator of the expression of enzymes from glycolytic and pentose phosphate pathways (PPP) and lipogenesis. Additionally, Akt-mediated activation of mTORC1 induces HIF-1α and sterol regulatory element-binding protein (SREBP) 1c. The SREBP1c transcription factor controls the expression of glucose-utilizing enzymes and enhances protein translation [215]. Primarily, HIF-1α upregulates the expression of genes encoding glycolytic enzymes and SREBP1c—the expression of PPP enzymes and of those involved in the *novo* lipid synthesis [218]. The expression of glycolytic enzymes is regulated also by c-Myc, which is upregulated by IL-7 as well [166]. In addition to glycolysis, c-Myc also controls glutamine metabolism [218], providing an alternative substrate for Krebs cycle and NADPH synthesis. Moreover, the IL-7/IL-7R signaling affects amino acid metabolism by controlling expression of their transporters Slc1a4, Slc1a5, Slc7a5, and Slc7a6 [219], as well as of the enzymes involved in amino acid synthesis [143]. Furthermore, the axis controls lipid metabolism by regulating glycerol uptake and triacylglycerol synthesis. It increases synthesis of aquaporin 9 [220], a glycerol membrane transporter, and stimulates the expression of enzymes involved in cholesterol synthesis [143].

2.6.4 IL-7/IL-7R Signaling in Cancer Invasion and Metastasis: The Association with Epithelial-to-Mesenchymal Transition (EMT) and Mesenchymal-to-Epithelial Transition (MET)

The EMT is a change of cell phenotype from epithelial to mesenchymal. It is involved in tumor progression, enabling invasion and metastasis, and in the generation of stem-like cells, facilitating resistance to anticancer treatment. The process involves the disruption of cell-cell and cell-matrix adhesion, cell cytoskeleton remodeling, and allows cancer cells to acquire migratory properties. At the molecular level, the EMT is characterized by the upregulated MMPs and a switch in gene expression, including decreased E-cadherin expression and elevated N-cadherin, fibronectin, and vimentin expression. The process involves transcription regulators ZEB, SNAI, and TWIST. The reversal route, MET, allows for establishing secondary tumors [221]. It is facilitated by heterogeneous nuclear ribonuclease H (hnRNPH), an enzyme involved in an alternative splicing and in suppressing the formation of the EMT-promoting ΔRON isoform [222].

The potential role of IL-7 in EMT/MET is unclear. Several lines of evidence seem to support an EMT-promoting role. In chondrocytes, IL-7 induces expression of S100A4, which promotes EMC degradation by upregulating MMP-13 via an autocrine loop involving receptor for advanced glycation end-products (RAGE) [129]. The S100A4, secreted also by fibroblasts, immune and cancer cells, is a known EMT inducer. It has been shown to downregulate

E-cadherin while upregulating MMPs, facilitating motility and adhesion of metastatic cancer cells [223]. The MMP-inducing activity is displayed also by oncostatin M, shown to be upregulated by IL-7/IL-7R axis as well [140]. Accordingly, IL-7 induces MMP-9 [54, 130], MMP-2 [130], MMP-3, and MMP-7 [47] in cancer cells. Moreover, in a breast cancer model, the IL-7/IL-7R axis facilitates the acquiring of mesenchymal phenotype by downregulating E-cadherin while upregulating N-cadherin [82]. Furthermore, IL-7 induces EMT markers ZEBs, TWIST1, SNAI, N-cadherin, and vimentin in prostate cancer cells and decreases E-cadherin. It also enhances the formation of tumor spheres, but, at least in this cancer type, has no effect on MMP expression [154]. However, contrary to the aforementioned observations, IL-7/IL-7R in hepatoma favors epithelial phenotype by upregulating E-cadherin and downregulating vimentin and α-SMA [54]. Whether the opposite effects on E-cadherin expression are associated with a cancer type or downstream pathways involved (PI3K/Akt and JNK in hepatoma [54] and PI3K/Akt in breast cancer [82])—need to be clarified. It might also be attributed to possible differences between canonical IL-7 [82] and its splicing variant, IL-7δ5, tested on hepatoma cells [54].

The IL-7 may be associated with EMT via Pim-1 or NFAT2 activation. The Pim-1 kinase has been shown to regulate, through eIF4B, the translation of c-Met and c-Met/hepatocyte growth factor (HGF) signaling, resulting in enhanced cell growth, survival, and motility. The upregulated activity of the pathway has been reported in a number of cancers and found to promote cancer cell growth, EMT, invasiveness, and metastasis [224]. The NFAT2 has been shown to downregulate the expression of E-cadherin in a TGFβ-independent manner, involving an upregulation of SNAI and ZEB transcriptional repressors. Moreover, NFAT2 has been demonstrated to induce the expression of Tks5, responsible for the formation of circumferential podosomes/invadopodia and cell-cell fusion and cancer invasion [163]. Furthermore, CD127 has been shown to be upregulated by ZEB2 in malignant lymphocytes [225]. On the other hand, however, IL-7/IL-7R is frequently shown to downregulate the expression of TGFβ, an EMT promoter, and to antagonize its signaling in cancer cells [56, 125, 226]. In fibroblasts, the interleukin has been observed to downregulate the expression of TGFβ [127, 128] and fibronectin [128]. It strengthens the cell-cell and cell-ECM adhesion by upregulating I-CAM expression on ovarian carcinoma [125] and melanoma cells [126], although others have failed to confirm the latter observation [227]. The recently reported phosphorylation of hnRNPH may potentially contribute to the MET rather than EMT [166].

The IL-7 induces cancer cell migration in esophageal squamous cell carcinoma [9], hepatoma [54], and breast [82], bladder [130], and prostate [47, 154] cancers. In turn, it has no effect on motility of lung cancer cells [6] and inhibits the migratory properties of fibroblasts [128]. However, the expression of CD127 in lung cancer correlates positively with tumor budding, a phenomenon of cancer cell dissociation from tumor invasive front [8].

2.6.5 IL-7/IL-7R Signaling in Cancer Angio- and Lymphangiogenesis

The IL-7 has been shown to induce VEGF-A expression in murine thymic cortical epithelial cells [228]. In gastric cancer, out of 52 analyzed cytokines, IL7 was the most strongly correlated with VEGF-A [229]. However, in vitro studies on human endothelial cell line HECV, displaying characteristics of vascular and lymphatic endothelium, do not confirm stimulatory effect of IL-7 on VEGF-A or B or on their receptors VEGFR1 and VEGFR2 [175]. The interleukin has been unable to stimulate VEGF-A and B in breast cancer cells as well [230]. Moreover, also in vivo, in murine cancer models, IL-7 does not increase the expression of vascular endothelial marker von Willebrand factor (vWF) and has no effect on microvessel density [81, 175]. On the contrary, immunotherapy with IL-7 decreases VEGF-A level at the tumor site while increasing an antiangiogenic CXCL10 (IP-10) in spontaneous murine bronchoalveolar cell carcinoma [231].

Instead, IL-7 has been shown to induce the expression of lymphatic markers in HECV, including podoplanin, prospero-related homeobox gene-1 (prox-1), and lymphatic vessel endothelial receptor-1 (LYVE-1). Moreover, the interleukin upregulates the expression of lymphangiogenic factor VEGF-D and its receptor VEGFR3. It also stimulates VEGF-C, albeit to a much lesser extent [175]. Furthermore, IL-7 enhances cell growth and migration of HECVs in a manner mediated by VEGF-D and facilitates microtubule formation in vitro and lymphatic tubule formation in vivo. As indicated by inhibition by Wortmannin, the lymphangiogenic activity of IL-7 involves PI3K, with probable downstream activation of HIF-1α [230]. The HECVs do not express IL-7, therefore the interleukin must be provided by other cells. Cancer cells, if expressing IL-7, stromal cells, dendritic cells, and/or macrophages are potential sources of IL-7 in the microenvironment. Al-Rawi et al. [46] demonstrated that breast cancer cells do not synthesize IL-7 but, upon IL-7 stimulation, express VEGF-D, which may subsequently bind to VEGFR3 on endothelial cells. Still, Iolyeva et al. [24] reported the presence of both IL-7 and its receptor in primary dermal human lymphatic endothelial cells. Ming et al. [6] and Jian et al. [81] showed VEGF-D to be a major downstream gene of IL-7 in lung cancer cells and found AP-1 (c-Fos/c-Jun) pathway to be involved [6]. The binding of AP-1 to VEGF-D promoter, triggered by IL-7, induces the formation of c-Fos and c-Jun heterodimers and enhances their DNA binding activity. In murine lung cancer models, IL-7 increases the formation of lymphatic tubules [81]. In clinical samples, both IL-7 and VEGF-D expression is associated with lymph node metastasis in non–small cell lung cancer [6].

2.6.6 IL-7/IL-7R Signaling and Autophagy

Only recently, IL-7 has been implicated in the inhibition of autophagy. Jian et al. [58] demonstrated that IL-7 stimulation of lung adenocarcinoma cells results in a drop in beclin-1 expression, a positive regulator of autophagy, frequently deleted or reduced in certain cancers. Downregulation of beclin-1 rescues PI3K/Akt/mTOR, which becomes phosphorylated and activated. The pathway leading from IL-7 to beclin-1 downregulation has not been explored. However, β-catenin is known to suppress beclin-1 expression [190] and, at least in MM, IL-7 activates β-catenin by upregulating MUC1 [188]. Moreover, in T-ALL cells, PI3K/Akt/mTOR pathway is utilized by IL-7 to induce phosphoribosomal protein S6, a known inhibitor of autophagy [172].

2.6.7 IL-7/IL-7R Signaling and the Effectiveness of Anticancer Therapy

The accounts regarding IL-7/IL-7R association with drug effectiveness are inconsistent. In murine model of colon cancer, the combination therapy of IL-7 and oxaliplatin, but not IL-7 alone, has resulted in reduced proliferation and increased apoptosis of cancer cells, lowering the number of abdominal and lung metastatic nodules. It has increased tumor infiltration with activated CD8+ T cells and reduced Treg count in the spleen, but had no effect on the counts of tumor-associated macrophages (TAMs), dendritic cells (TADCs), or myeloid-derived suppressor cells (MDSCs) [232]. In a line with a sensitizing role, the CD127 expression is reduced in pancreatic cancer cell lines resistant to the three typical chemotherapeutics [233]. Others, however, have demonstrated that a gain of IL-7 was one of four unique aberrations associated with cisplatin resistance in a human glioma cell line, conferring protection against cisplatin-induced apoptosis [4]. Moreover, the overexpression of mutated CD127 induces resistance to steroids in T-ALLs. Mechanistically, steroid-resistance is associated with a strong activation of Akt and MEK/ERK pathways and transducers downstream of Akt, that is, p70-S6K, CREB, and NF-κB and with inhibition of GSK3β, resulting in an upregulated expression of Mcl-1 and Bcl-x$_L$, and with MEK/ERK-mediated downregulation of Bim [173].

The ability of IL-7 to confer protection against steroid-induced apoptosis was corroborated by Wuchter et al. [204] in established T-ALL cell lines, as well as in patient-derived primary cultures. It was also observed by Delgado-Martin et al. [150] in early T-cell precursor (ETP)-ALL and non-ETP-ALL. In addition, IL-7 treatment antagonized HDAC inhibitor in ETP-ALL [150]. Moreover, IL-7 partially protects the malignant Hodgkin-Reed-Sternberg cells from doxorubicin-induced apoptosis [57]. The interleukin rescues also T-ALL primary cells while it is mostly ineffective in established cell lines. Two exceptions have been noted, in which IL-7 either inhibited apoptosis (KE-37 line) or, contrary, enhanced cell sensitivity to doxorubicin (MOLT-3 line), thus displaying a proapoptotic activity [204]. In chronic myeloid leukemia (CML), IL-7 derived from myeloid stem cells protects cancer cells from apoptosis induced by inhibitors of tyrosine kinases, imatinib or nilotinib, in a JAK1/STAT5-dependent manner [34, 149]. In preB-ALL and BCP-ALL, it protects malignant cells from rapamycin-induced cell death [181, 211].

2.6.8 Immunomodulatory Role of IL-7/IL-7R in Tumor Microenvironment

It has been repeatedly observed that IL7-transfected cancer cells display reduced tumorigenicity and alleviate tumor-induced immunosuppression when grafted into mice [56, 226, 227, 234–238], giving rationale for IL-7 application in anticancer immunotherapy of solid tumors. Using various delivery systems, IL-7 has been shown to effectively reduce tumor burden, either as a single immunomodulator or in combination, either in its native or modified form (reviewed in [16, 17, 239]). The IL-7 is particularly attractive as a therapeutic because it does not induce hyperinflammation. Moreover, it expands naïve and memory T cells selectively, without inducing proliferation of immunosuppressive Tregs. Apart from providing mitotic and survival stimuli, IL-7 enhances the cytotoxicity of cancer-specific cytotoxic T lymphocytes

(CTLs), monocytes, NKT, NK, and LAK cells. It induces perforin secretion by CTLs in a STAT5-dependent manner [147] and stimulates expression of IFNγ [203, 231, 240–242], IL-12, MIG, and IP-10 [203, 231]. In addition, IL-7 enhances IL2 mediated LAK activity [240] and induces IL-2R expression on T cells, potentially improving their responsiveness to IL-2 [243, 244]. Upon IL-7 stimulation, human peripheral monocytes release a number of cytokines including IL-6, IL-1α, IL-1β, TNFα, and MIP-1β, and their tumoricidal activity toward human melanoma cell line is enhanced [245, 246]. The TNFα production is enhanced also in PBMCs [112]. The IFNγ, in the presence of TNFα, promotes polarization of macrophages into M1 phenotype, displaying proinflammatory and antitumor properties [247]. Accordingly, IL-7 has promoted M1-resembling phenotype in tumor macrophages in lung cancer model. The cells were characterized by upregulated expression of IL-12 and inducible nitric oxide synthase (iNOS) and downregulated IL-10 and arginase [248]. Moreover, the interleukin has been shown to induce secretion of proinflammatory cytokines IL-1β, IL-6, and TNFα, IFNγ, chemokines IP-10, MIP-1α and β, and RANTES from cervicovaginal epithelial cells. The IL-7 has also been reported to upregulate the expression of adhesion molecule V-CAM on endothelial cells, facilitating the recruitment of immune cells from circulation [249]. Proinflammatory environment is necessary for differentiation, maturation, and efficient activation of antigen presenting cells (APC) and thus for their ability to provide costimulatory signals for cytotoxic T cells. Host APCs are necessary for successful antigen presentation even in case of highly immunogenic tumors [250]. Furthermore, IL-7 not only increases the counts of CD4+ and CD8+ T cells, diminished at the tumor site [16, 242], but also enriches CD19+ B-cell population. This activity is unique for IL-7 and potentially facilitates antibody-dependent cell-mediated cytotoxicity [251]. The interleukin prompts the CD4+ T-cell differentiation into Th9 cells and increases their antitumor activity, dependent on IL7-induced, upregulated expression and secretion of IL-9 and

IL-21 [152]. It may also impart polyfunctionality and stemness on CD4+ T cells in a STAT5-dependent manner, involving increased expression of EZH2, a histone methyltransferase. Subsequent histone H3 acetylation and increased chromatin accessibility in the promoter regions of IFNγ, IL-2, TNFα, and granzyme B allow for their concomitant synthesis. The ability to mediate diverse effector functions is a characteristic feature of polyfunctional phenotype. The interleukin-induced polyfunctional CD4+ T cells have been further characterized by low expression of PD-1 and Foxp3 markers and IL-17A [151]. The IL-7 may also enhance cytolytic function of immune cells by increasing mRNA and membrane expression of FasL, as has been demonstrated for NK cells [206] and CTLs [205]. In addition, the interleukin provides costimulatory signal for the interaction between cytotoxic lymphocytes and APC by increasing lymphocyte expression of ICOS [252].

Immunosuppressive character of tumor microenvironment is associated with predominance of TAMs with M2 polarization characteristics, Tregs and Th2, MDSCs, and CAFs, which, supported by cancer cells, secrete immunoinhibitory IL-6, IL-10, and TGFβ. They block the functionality of CTLs and dendritic and NK cells, while further promoting trafficking and expansion of tumor-tolerant immune cells [16, 247]. In addition to CD4+ and CD8+ T cells, IL-7 treatment enriches tumor bed with Th1 (CD4+IFNγ+) and Tc1 (CD8+IFNγ+) T cells [242], basophils and eosinophils [237], γδT cells, and with conventional and monocyte-derived DCs [249]. The IL-7 inhibits TGFβ production in fibroblasts [134], macrophages [253], and tumors [56, 125, 203, 226, 231], engaging, at least in fibroblasts, the JAK1/STAT1 pathway [134]. Additionally, IL-7 interferes with TGFβ signaling by inducing SMAD7, a TGFβ suppressor [127, 128, 203, 242]. The SMAD7 mediates IL-7 inhibitory effect on TGFβ fibrotic activity via downregulating PCKδ phosphorylation in fibroblasts [127, 128]. It is also involved in the inhibition of Foxp3 expression, which is a critical factor for the development and function of Tregs [203, 242]. Moreover, IL-7 decreases the production of other mediators of immunosuppression in tumor and spleen, including IL-10, prostaglandin E2, and VEGF-A [203, 231, 242, 249].

However, IL-7 effect might depend on the context and the environment. Sin et al. [254] confirmed IL-7 ability to enhance the cytolytic activity of CTLs. Yet, the authors have shown that the cytokine eluded the classic Th1 and Th2 paradigm. In splenocytes, it decreased the expression of IL-2 and IFNγ, classic Th1 cytokines, but increased that of IL-10, a classic Th2 cytokine. It also downregulates RANTES while upregulating MCP-1, a feature consistent with Th2 cytokine behavior [254]. Moreover, some of immunomodulatory IL-7 activities might be viewed as immunosuppressive and tumor-promoting. The interleukin stimulates IL-6 production in monocytes [245], bone marrow stroma [61], and CAFs [57]. Lactate, a product of IL-7-accelerated glycolysis [155, 180, 216], is a metabolic trigger of immunosuppression, known to inhibit CTLs' activation and promote metastasis [255]. The role of ICAM1, upregulated on cancer cells by IL-7 [125, 126], is context-dependent. It may allow for T-cell attachment to the tumor cell via LFA-1 receptor and enable a proper orientation of cytotoxic attack. Still, it may also benefit tumor, if ICAM binds to LFA-1 receptor on macrophages, initiating spheroid formation and facilitating metastasis [256]. Furthermore, IL-7 promotes the expression of programmed cell death 1 (PD-1), cell surface mediator of immunosuppression [255], on PBMCs and on purified CD4+ and CD8+ T cells (memory, memory-effector and effector subsets, but not on naive T cells). In addition, it induces PD-L1, a PD-1 ligand, on T cells and monocytes/macrophages [257]. Tumors frequently express PD-L1 and exploit the PD-1/PD-L1 interaction to evade immunosurveillance. The consequence of PD-1/PD-L1 engagement is SMAD3-mediated differentiation of naïve T cells into Tregs, with concomitant inhibition of CTLs expansion and cytotoxicity. It is accompanied also by a reversed metabolic switch—from lactate-generating glycolysis to β-oxidation and oxidative phosphorylation, resembling the exhausted T-cell phenotype [258]. Kinter et al. [257] showed that in case of IL7-stimulated PD-1

expression, PD-L1 did not affect IL7-induced T-cell proliferation, survival or STAT5-mediated functionalities but significantly reduced the interleukin-dependent cytotoxicity. The impact on metabolism has not been explored. Still, Myklebust et al. [259] reported signaling deficits in STAT5 phosphorylation induced by IL-7 in PD-1[high] CD4+ T cells infiltrating follicular lymphoma tumors. Interestingly, high PD-1 expression on human tumor TILs has been linked with lack of CD127 [260]. Consistently, the focal point of anti-PD-1 therapy aimed at rejuvenating CD8+ T cell is the upregulation of CD127 expression on exhausted cells [261].

Other tumor-promoting immunomodulating activities of IL-7 have recently been discovered. IL-7 preferentially stimulates proliferation of IL17-producing subsets of γδT cells [75, 138, 262], known for their proangiogenic and MDSCs-recruiting activity. Chen et al. [262] showed that the frequency of those cells, as well as number of co-localized *IL7* transcripts, is significantly higher in lymph nodes from aged mice. Moreover, they have found that, upon tumor challenge, IL17-producing γδT cells migrated to the tumor bed and contributed to the immunosuppressive environment, facilitating growth of larger tumors than those developed in young mice. In addition, others have shown that IL-7 induced the expression of CD47, a "do not eat me" signal, on cancer cells, thus enabling evasion of host phagocytes [263]. An overview of immunomodulatory role of IL-7 in tumor microenvironment is depicted in Fig. 2.2.

2.7 IL-7/IL-7R Signaling and Cancer Type

2.7.1 Hematological Malignancies

The IL-7/IL-7R has been implicated in the development of several human hematological malignancies, namely, T- and B acute lymphoblastic leukemias, cutaneous T-cell lymphoma, multiple myeloma, chronic myeloid leukemia, and classic Hodgkin's lymphoma. The IL-7/IL-7R signaling has been the most extensively studied in

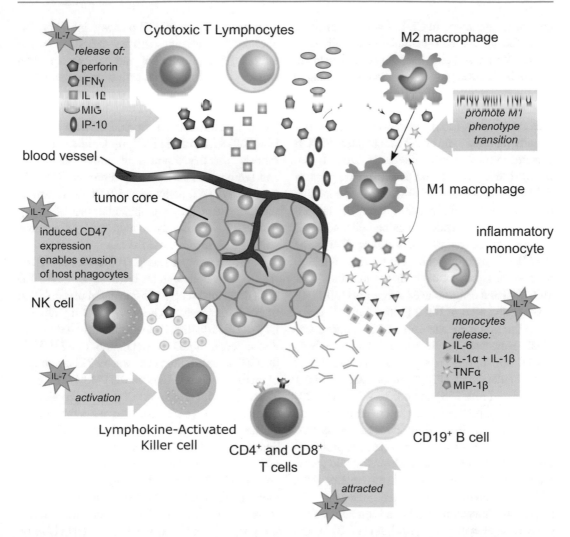

Fig. 2.2 Immunomodulatory role of IL-7 in tumor microenvironment. IL-7 elicits several processes in the microenvironment of a tumor. IL-7 activates natural killer cells (NK) and lymphokine-activated killer (LAK) cells. Stimulation with IL-7 leads to increased release of several cytokines such as perforin, interferon γ, IL-12, MIG, IP-10 cytotoxic T lymphocytes. Inflammatory monocytes release tumoricidal IL-6, IL-1α, IL-1β, TNFα, and MIP-1β. IL-7 attracts B cells and CD4+ / CD8+ T cells to the tumor's microenvironment. IL-7 induces expression and presentation of the CD47, a "do not eat me" signal, thus enabling evasion of host phagocytes. IL-7 stimulates release of IFNγ and TNFα, which in turn promote transition of macrophages to M1 phenotype

T-ALL. As mentioned earlier, the responsiveness of malignant lymphocytes to IL-7 depends on the CD127 expression [76, 78]. Neoplastic cells of the T- or B-phenotype and of myeloid lineage, from chronic as well as from acute leukemias, proliferate in response to IL-7 in a dose-dependent manner [209]. While mostly unable to overcome a differentiation block in malignant lymphocytes [264], IL-7 stimulation has sporadically resulted in decreased DNA synthesis accompanied by expression of differential markers [265].

2.7.1.1 T-Cell Acute Lymphoblastic Leukemia (T-ALL)

In T-ALL, malignant T cells express high levels of CD127 [209, 266, 267], while the interleukin is released by thymic epithelial cells [30] and bone marrow stroma [33]. The IL-7/IL-7 signaling plays a crucial role in T-ALL by being a transforming, mitogenic, antiapoptotic, and trophic factor. It has been shown that IL-7 transgene is enough to promote neoplastic transformation [26], which can also occur in the presence of combined mutations in CD127 and N-Ras genes [84]. Except for gain-of-function mutations in CD127 [89], also genes encoding proteins downstream IL-7R are frequently mutated, resulting in the hyperactivity of IL-7/IL-7R pathway, with the strongest level of activation conferred by JAK1 mutations [268]. Moreover, the most frequently mutated gene in T-ALL encodes NOTCH protein, known to transactivate IL-7/IL-7R signaling [269, 270]. In addition, the frequent in ETP-ALL mutations in *DNM2* gene, encoding cytoskeleton protein dynamin 2, impair clathrin-mediated CD127 endocytosis and enhance cell-surface expression of the receptor [271]. The IL-7/IL-7R signaling enhances survival of malignant T cells preventing both spontaneous [196, 197, 272] and drug-induced apoptosis [150, 204]. The upregulation of Bcl-2 [140, 196, 197] seems to be the major mechanism of IL7-induced survival. However, the interleukin has also been shown to slightly upregulate Bax [196] and downregulate p53 [172]. An increase in Mcl-1 and Bcl-x_L [140, 173] and decrease in Bim [173] have been implicated in the interleukin-mediated protection against drug-induced apoptosis. Activation of IL-7/IL-7R axis induces proliferation of malignant cells and allows for colony formation by facilitating cell cycle progression [197, 266]. Mechanistically, IL-7 triggers Rb hyperphosphorylation [197] and upregulates the expression of cyclins D2 [197], A [197, 272], and E [272]. It also increases the expression and activation of cdk 4 and cdk2 [197] and downregulates p21^{CIP1} [197] and p27^{KIP1}, additionally interfering with its stability [140, 172, 197, 272]. The interleukin-mediated maintenance of mitochondrial membrane potential, crucial for cell viability, is dependent on intracellular ROS. The ROS generated by ETC are required for PI3K/Akt/mTOR activation, which, subsequently, promotes continuous upregulation of ROS production by accelerating glucose uptake [217]. The IL-7 relevance for T-ALL survival, proliferation, and viability was corroborated in vivo by Silva et al. [198]. The IL-7/IL-7R signaling via STAT5 has been implicated in inducing differentiation arrest in T-ALL cells by activating AUTS2, with subsequent expression of homeobox protein MSX1 [144].

Rapamycin, an mTOR pathway inhibitor, has abrogated mitogenic activity of IL-7, while affecting the prosurvival one only to a certain degree [197]. Still, PI3K activation is mandatory for the interleukin-induced upregulation of Bcl-2 and downregulation of p27^{KIP1}, hyperphosphorylation of Rb, and for trophic activity of IL-7 toward T-ALL cells [172, 180]. Unlike in murine models, the development of human leukemias and lymphomas seem to rely more on PI3K/Akt than JAK/STAT pathway [180]. This, however, may depend on the developmental stage of malignant cells. Accordingly, Maude et al. [78] demonstrated that in all screened cases of ETP-ALL, and only a few of non-ETP T-ALL, the IL-7/IL-7R activated JAK1/STAT5. The ETP-ALL cells depended on STAT5-mediated upregulation of Bcl-2 for survival. The IL-7/IL-7R signaling in ETP-ALL is distinct from that in non-ETP T-ALL also in terms of CD127 regulation. In ETP-ALL, the CD127 expression is driven by ZEB2, a target of activating mutations in ETP-ALL. In turn, ZEB2 has no effect on CD127 expression in mature T-ALL [225]. However, moderating their earlier conclusions [171], Robeiro et al. [140] demonstrated that PI3K/Akt activation is responsible for IL7-induced cell cycle progression and cell growth, but cannot fully replace IL-7 in promoting cell viability. The authors have shown the requirement also for STAT5 in providing mitogenic signals and ensuring T-ALL cells viability. Yet, they have also noted the independence of Bcl-2 upregulation from STAT5 activation. In addition, Robeiro et al. [140] reported STAT5-mediated activation of Pim1 kinase, which partake in IL7-induced T-ALLs proliferation. Apart from Pim1, IL-7 induced STAT5-mediated upregulation of genes encoding inhibitors of IL-7/IL-7R signaling pathway (CISH, SOCS2) and a CD127 repressor (Ikaros). It has also downregulated BCL6 and IL-10.

The MEK/ERK pathway is activated by IL-7 in T-ALL, but its relevance is not clear. However, it has been implicated in drug resistance [173] and protein translation [182]. Ksionda et al. [176] demonstrated that IL-7 might contribute to the survival and proliferation of leukemic blasts by activating a non-canonical RasGRP1-Ras/PI3K/Akt pathway.

Melao et al. [272] showed that in T-ALL, IL7-mediated cell viability depends on the interleukin activation of casein kinase 2 (CK2), which regulates both PI3K/Akt, via deactivating PTEN, and JAK/STAT pathways. The CK2 seems to be crucial for antiapoptotic (Bcl-2 upregulation), mitogenic (cyclin E and A upregulation and p27^{KIP1} downregulation), and trophic activity of IL-7 (maintaining cell size and mitochondrial potential). The CK2 has mediated IL-7 activity in case of both wild type and mutant CD127 receptor.

2.7.1.2 Cutaneous T-Cell Lymphoma (CTCL)

Normal and CTCL-derived skin fragments have detectable expression of CD127 and IL-7, more prominent in cancer samples [60, 128]. The CD127 is present in basal keratinocytes and in skin-infiltrating lymphocytes [60], while IL-7 in keratinocytes [128, 273] and fibroblasts [128]. The CTCL-derived skin explants have induced the interleukin-mediated T-cell proliferation while normal skin has not [128]. Other study has pointed at keratinocytes as the IL7-producing cells, capable of supporting the growth of cancer cells [273]. Qin et al. showed IL-7 to stimulate DNA binding of JunD, which facilitated cell transition from G1 to S phase. The authors have hypothesized that the activation of c-Myb and cyclin D1, known JunD transcriptional targets, might mediate the effect [207]. Apart from mitogenic stimuli, IL-7 provided survival signals. It has been demonstrated that the cytokine upregulated Bcl-2 expression, both by enhancing STAT5 binding to Bcl-2 promoter and by activating c-Myb [141]. Recently, IL-7 has been shown to facilitate CTCL cells evasion of immune surveillance by inducing cell surface expression of CD47, a "do not eat me" signal for phagocytes [263].

2.7.1.3 B-Cell Acute Lymphoblastic Leukemia (B-ALL)

The IL-7/IL-7R signaling participates in survival, proliferation, and differentiation of B-cell progenitors (reviewed in [139]). Its transforming activity has been demonstrated as well [27]. In B-ALL, but not T-ALL, CD127 expression is upregulated by the mutations in *IKZF1* gene, encoding for Ikaros, a negative regulator of CD127 transcription [274]. Receptor expression on malignant B cells depends on the disease stage [275] and is associated with widespread nodal dissemination [276]. Patients with pre B ALL, but not those with B-CLL or B-NHL, have higher frequency of CD127$^+$ B-cell blasts than the frequency of CD127$^+$ CD19$^+$ PBMCs in healthy individuals. Moreover, CD127$^+$ preB-ALL cells have a greater proportion of Ki-67$^+$Bcl-2^{high} cells compared with CD127$^-$ counterparts, implying the involvement of IL-7/IL-7R signaling in B-ALL proliferation and survival [86]. In BCP-ALL, the impact of IL-7 on proliferation is mediated by JAK/STAT1 and STAT5, although the formation of active STAT3 upon the interleukin stimulation has been also observed [135]. The IL-7 ensures proliferation and survival of preB-ALLs and can rescue those cells from rapamycin-induced apoptosis [211]. Recently, a cross-talk between IL-7/IL-7R signaling and Y-box-binding protein 1 (YB-1), a multi-drug resistance marker, has been reported. The interleukin induced the phosphorylation of YB-1 via MEK/ERK/RSK2 and mediated, in part, survival activity of IL-7 in rescuing BCP-ALL cells from rapamycin-induced apoptosis [181].

2.7.1.4 Multiple Myeloma (MM)

The MM develops mostly in bone marrow, destructing the bone structure by enhancing osteoclastogenesis and inhibiting osteoblastogenesis. The IL-7/IL-7R signaling has been implicated in both. Stromal bone marrow cells produce both IL-7 and CD127 [61], and bone marrow plasma from MM patients contains higher IL-7 concentrations than recorded in normal bone marrow [36]. In response to monocyte-derived proinflammatory IL-1β and TNFα, stromal cells and osteoblasts synthesize IL-7, which stimulates T cells to release osteoclastogenic factors—receptor activator of NF-κB ligand (RANKL) and macrophage colony-stimulating factor (M-CSF) [36, 277]. It also reduces synthesis of osteoprotegrin [277], a RANKL decoy receptor, by inhibiting Runx2/Cbfa1 activity in human preosteoblasts [277,

278]. Others have shown that IL-7 can be also produced by MM cells [279], and MM-derived IL-7 induces the expression of Gfi1, a Runx2 suppressor, in preosteoblasts. The Gif1-mediated Runx2 suppression involves the HDAC and blocks cell differentiation [279]. Hiasa et al. [158] implicated Pim2 kinase in IL7-mediated osteoclastogenesis. Mechanistically, Pim2 inhibits BMP2-mediated upregulation of Osterix expression by antagonizing SMAD1/5 and p38MAPK signaling. In turn, the participation of HIF-1α activation in the stimulation of IL-7 expression has been suggested [280]. In MM cells, but not normal B-cells, IL-7 induces MUC1 expression and enhances its binding with β-catenin; however, the relevance for MUC1 induction in MM has not been determined [188].

Apart from MM, the IL-7/IL-7R signaling has been implicated in the spontaneous osteoclastogenesis occurring in cancer patients with osteolytic metastasis. However, the IL-7 involvement seems to relay rather on enhancing the activity of TNFα, as there is no interleukin-mediated change in RANKL [112].

2.7.1.5 Classic Hodgkin's Lymphoma (cHL)

Both IL-7 and CD127 are expressed by malignant cells in cHL and elevated IL-7 concentrations, corresponding with the disease advancement, have been found in patients' sera. The IL-7/IL-7-R signaling, in an autocrine and/or paracrine manner, is believed to provide a growth and survival support for malignant Hodgkin and Reed-Sternberg (H-RS) cells (reviewed in [281]). The cHL-derived CAFs not only secrete IL-7 and induce clonogenic growth of cHL cells, but also release IL-6, an important growth factor in cHL and a mediator of immune tolerance. As bone marrow stromal cells express CD127 [61], IL-7 synthesized by malignant cHL cells has been shown to stimulate IL-6 production in CAFs [57].

2.7.1.6 Chronic Myeloid Leukemia (CML)

The CML originates from pluripotent hematopoietic stem cells, in which CML-specific CTLs play a crucial role in immunosurveillance, preventing the disease progression to the terminal blast phase. The CML-specific CTLs express high levels of CD127 and the granulocyte-derived IL-7 expression in spleen is higher in CML than normal mice, implying the role of IL-7/IL-7R signaling in the CD8+ T-cell-mediated immunosurveillance [43]. However, in the blastic crisis phase, MSCs-derived IL-7 seems to play an adverse role, rendering malignant cells resistant to imatinib or nilotinib, tyrosine kinase inhibitors. Antiapoptotic activity of IL-7 is dependent on JAK1/STAT5 pathway and does not involve JAK3/STAT5 or PI3K/Akt [34, 149].

2.7.2 Solid Tumors

As a part of cancer immunosuppression, stromal cells in lymphoid organs of individuals with solid tumors may be reprogramed to downregulate IL-7/IL-7R activity. Indeed, Riedel et al. [282] showed that in melanoma-inoculated mice, some unidentified tumor-released factors altered transcriptional patterns of tumor-draining lymph nodes reducing IL-7 expression in fibroblastic reticular cells. Consequently, consistently with IL-7 function, the proliferation of T and B cells was reduced and the alterations in node immune composition occurred [282]. Others have shown that IL-7 might participate in thymic involution observed in tumor bearers. Experimental animals with mammary tumors had a shrunken thymus and downregulated thymic IL-7 expression. The organ volume could be partially restored by intrathymic injection of recombinant interleukin [283]. Moreover, Mandal et al. [284] demonstrated that developing Ehrlich's ascites carcinoma (EAC) results in depletion of T-cell subpopulations in the thymus. Mechanistically, an unknown EAC-derived soluble factor induced CD127 downregulation, inhibited JAK3 and STAT5 phosphorylation, and decreased Bcl-2/Bax ratio, resulting in thymocyte apoptosis and T-cell maturation block. As a result, the thymus atrophied, and T-cell subpopulations among PBMCs were depleted. On the other hand, however, a body of evidence is being gathered showing that IL-7 might be upregulated in cancer, both at the systemic and local level [2–10, 100–112], and support the growth of solid tumors. In the following paragraphs, what little is known about IL-7/IL-7R signaling in particular cancer types is presented.

2.7.2.1 Prostate Cancer

The IL-7 and its receptor are present at mRNA and protein level in normal prostatic epithelia and in endothelial cells, allowing for autocrine and paracrine mode of signaling. The CD127 has also been detected on intraepithelial and parafollicular T cells and on HeV-like vessels. The interleukin presence in prostate seems to support T-cell survival rather than proliferation as has been evidenced by low proliferation rates but high Bcl-2 expression [44]. However, neoplastic transformation dramatically decreases IL-7 content [44, 285]. The phenomenon has been accompanied by diminished frequency of CD8+ T cells positive for Bcl-2, implying that IL-7 in prostate cancer plays an immunosurveillance function [44]. Correspondingly, lower expression of IL-7 and IL-7R is predictive of worse survival in prostate cancer patients [286], and IL-7 vaccination improves survival of animals challenged with prostatic cancer cells. Mechanistically, IL-7 increases intratumoral necrosis by enhancing tumor infiltration with CD4+ and CD8+ T cells and with CD19+ B cells [251]. Others, however, have found IL-7 to be elevated in neoplastic prostate tissue compared to benign hyperplasia already in stage I cancers [5]. The Gene Set Enrichment Analysis (GSEA) of 551 prostate cancer samples has confirmed IL-7 and CD127 co-expression and showed their high positive correlation. Moreover, the GSEA has revealed that in patients stratified by the degree of IL-7/IL-7R expression, those in the highest decile had upregulated expression of genes associated with stemness and metastasis. Therefore, a positive correlation between the expression of IL-7/IL-7R and metastasis and EMT in prostate cancer has been suggested. In addition, there has also been a positive correlation with IL-6/STAT3, TNF, NF-κB, EGF/PDGF, and Wnt signaling pathways [154]. The interleukin and its receptor seem to be

expressed only by androgen receptor (AR)-negative lines [5, 154], although others have reported their presence also in LNCaP, an AR-positive cell line [47, 70]. Saha et al. [70] showed STAT3/NF-κB to be involved in the upregulation of IL7 transcripts. The IL-7/IL-7R signaling activated Akt/NF-κB pathway [47] and STAT5, Akt, and ERK [154]. The in vitro data point at IL-7/IL-7R involvement in cell migration and invasion [47, 154]. However, while Qu et al. [47] demonstrated the involvement of MMP-3 and MMP-7, upregulated by the interleukin, Seol et al. [154] failed to confirm IL-7 effect on *MMP1*, *MMP2*, *MMP7*, *MMP9*, and *MMP13* transcripts or MMP-2 and MMP-9 activities. Instead, the authors have shown IL-7 to upregulate the mRNA and/or protein of EMT-promoting transcription factors ZEBs, TWIST1, and SNAI as well as to promote the switch from epithelial (E-cadherin) to mesenchymal markers (N-cadherin, vimentin). Moreover, they have demonstrated that IL-7 stimulation induced sphere formation in cancer cells [154]. The interleukin association with metastasis in prostate cancer has been also manifested in clinical setting by more markedly increased cytokine concentrations in patients with bone metastases [112].

2.7.2.2 Fibrosarcoma

The only available data concerning IL-7 association with fibrosarcoma are those regarding inhibitory effect exerted on tumors in mice by recombinant IL-7 or by a genetically modified IL7-expressing fibrosarcoma cell lines. The interleukin reduces tumor growth rates or causes their complete rejection. It also confers a resistance to re-challenge, contributes to less severe inhibition of lymphocyte proliferation with improved the CD4+:CD8+ ratio and cytotoxicity, and reduces synthesis of TGFβ [226, 237].

2.7.2.3 Esophageal Cancer

Tumor and tumor-adjacent normal tissue express IL-7 at both mRNA [287] and protein level [2], but only the interleukin protein content is higher in tumors. Tumor-to-normal ratio of IL-7 is increased in advanced cancers and those metastasizing to lymph nodes, due to decreased expression in tumor-adjacent tissue [2]. The mRNA for IL-7 has been detectable in five out of six screened cell lines [288]. The CD127 is overexpressed in esophageal squamous cell carcinoma (ESCC) and esophageal adenocarcinoma as compared to normal esophagus as well and correlates with shorter disease-free survival [9]. Its expression in ESCC cell lines is controlled by FoxO1, which, if acetylated, acts as CD127 transcriptional repressor. In cancer, FoxO1 is deacetylated due to HDAC overexpression. The IL-7 receptor in ESCC acts as an oncogene. Its knockout results in diminished cell migration and brief G2 cell cycle arrest, accompanied by p21^{CIP1} upregulation [9].

2.7.2.4 Gastric Cancer

The IL-7 concentration in gastric tumors is higher than in surrounding macroscopically normal tissue. Cancer advancement, particularly lymph node metastasis, is associated with a bigger difference in expression ratio between tumor and normal tissue. However, it results from a drop in IL-7 concentration in normal tissue in advanced and lymph node positive cancers [2]. The IL-7 mRNA has been detected in primary tumors of gastric cancer [287, 289] but not in bone metastases [289]. Still, early studies have not shown cytokine expression in two screened gastric cell lines [71]. No studies concerning IL-7 signaling in gastric cancer seem to have been conducted, although CD127 expression has been shown in patient samples both at mRNA and protein levels [10, 287], positively reflecting lymph node involvement and cell dedifferentiation and predictive of poor prognosis and cancer recurrence [10]. Moreover, *Helicobacter pylori* infection, a risk factor for gastric cancer, is accompanied by elevated *IL7* mRNA, proportionally to the severity of infection [290]. A genome-wide association study (GWAS) has identified 16 genes from IL-7 pathway as associated with increased susceptibility to gastric cancer, of which NMI, an N-Myc and STAT interactor, has been upregulated in tumors as compared to normal tissue [291].

2.7.2.5 Colorectal Cancer (CRC)

Interleukin concentrations in the colorectum, both in tumor and macroscopically normal tumor-adjacent tissue, are significantly higher than in the upper digestive tract. Like in esophageal and gastric cancer, IL-7 concentration in colorectal tissue of CRC patients is affected by cancer stage only in tumor-adjacent tissue (decreased in advanced and lymph node metastatic cancers) [2]. The *IL7* transcripts have been detected in clinical samples, both in tumor and adjacent normal tissue, as well as in three out of four established cell lines [80]. Gou et al. [232] demonstrated that IL-7 enhanced antitumor activity of oxaliplatin by inhibiting the growth of lung and abdomen metastases in murine cancer model, which was associated with increased tumor infiltration with activated CD8$^+$ T cells and reduced Tregs content in the spleen.

2.7.2.6 Lung Cancer

Lung cancer cells overexpress CD127 [202], and the receptor expression in stage I cancers is positively correlated with tumor budding [8], shorter recurrence-free and overall-survival, and with CD68$^+$, a TAM marker [7, 58]. High IL-7 and CD127 expression has been associated with aggressive cancer phenotype: advanced stage, lymph node metastasis, VEGF-D expression, and higher lymphatic vessels density [6]. A protumorigenic activity of IL-7 in lung cancer is confirmed by in vitro findings. The IL-7/IL-7R has induced the expression of c-Fos, expression and phosphorylation of c-Jun, and enhanced AP-1 binding with DNA. It has also upregulated the expression of cyclin D, resulting in cell cycle progression and accelerated proliferation. Moreover, it has also increased expression of VEGF-D, without affecting VEGF-A and VEGF-C, and induced formation of lymphatic vessels and tumor growth in vivo [6, 81, 210]. The IL-7/IL-7R axis seems to have no effect on cancer cell migration [81], but it has improved cell survival by activating p53 and inducing Bcl-2 expression while downregulating Bax [202]. The pathway in lung cancer has being linked with prevention of autophagy as it decreased beclin-1 expression, a positive autophagy regulator. Interestingly, it resulted also in the downstream activation of PI3K/Akt/mTOR pathway [58].

2.7.2.7 Renal Cell Carcinoma (RCC)

Renal cell carcinoma cells express and secrete IL-7, more so in response to IFNγ stimulation [71]. The interleukin, in turn, can induce IFNγ secretion from RCC-reactive TILs and potentiate their growth. It has also been shown to enhance IL2-mediated LAK activity, albeit to lesser extent [240]. Antitumor activity of IL-7 in RCC has been also demonstrated by Ditonno et al. [243] and associated with enhanced proliferation of TILs, upregulated expression of IL-2R, and induction of phenotypical changes in lymphocyte populations with CD4+ T-cell dominance.

2.7.2.8 Bladder Cancer

The effects of IL-7/IL-7R signaling in bladder cancer have been analyzed on 5637 cell line by Park et al. [130], who demonstrated an induction of wound healing migration and invasion but no effect on proliferation. Upon IL-7 stimulation, the phosphorylation of ERK1/2, but not JNK and p38MAPK, has been observed. In line with the lack of mitogenic activity, IL-7 has not affected cyclin D1 or E, CDK2 or CDK4, p21^{CIP1} or p53. However, contrary to all other reports, IL-7 induced p27^{KIT1} expression. Moreover, p27^{KIT1} seemed to be required for ERK phosphorylation, as well as for subsequent enhanced binding of NF-κB and AP-1 to DNA, an upregulation of MMP-2 and MMP-9, and, finally, for cell migration and invasion.

2.7.2.9 Gynecological Cancers

In cervical cancer, an immunomodulatory role of IL-7, as an Fc-fused (IL7-Fc) interleukin to enhance its delivery across the genital epithelial barrier, has been examined [249]. The IL7-Fc has been successful in suppressing the growth of cervicovaginal tumors in mice. It has been shown to induce secretion of proinflammatory cytokines IL-1β, IL-6, and TNFα, IFNγ, chemokines CXCL10 (IP-10), CCL3 (MIP-1α), CCL4 (MIP-1β), and CCL5 (RANTES) from cervicovaginal epithelial cells, and stimulate endothelial V-CAM expression. The IL7-Fc has also enhanced infiltration of cervicovaginal tissue with CD4+ and CD8+, but their location differed. The CD4+cells preferentially localized in the submucosa and CD8+ in the epithelium. The populations of γδ T cells and of conventional and monocyte-derived DCs increased as well [249]. Others have examined lnc-IL7R and found it to be upregulated and correlate positively with tumor size, cancer stage, and lymph node metastasis and to be indicative of poor prognosis and shorter overall survival. Mechanistically, lnc-IL7R has prevented apoptosis by upregulating Bcl-2 expression in tumor cells and induced tumor growth in vivo [96].

No IL-7 expression has been detected in epithelial ovarian cancer cell lines [125], although in peritoneal fluid as well as at the systemic level, IL-7 has been elevated and indicative of shorter both disease-free and overall survival [105, 125]. Compared to plasma, IL-7 concentrations in cancer-associated ascites are decreased, and the ascites contain low count of CTLs. In addition, the ascites' CTLs display a limited antitumor response [292].

Among genetic alterations in endometrial cancer, the targeted next-generation sequencing (NGS) has revealed a high frequency of mutations in PTEN (50%), as well as mutations in genes encoding Akt2, PIK3R1, FoxO, RICTOR, MTOR, RPTOR, and TSC1. Taken together, those mutations translate into enhanced activity of IL-7/IL-7R signaling [293].

2.7.2.10 Melanoma

Melanoma cells have been reported to express both IL-7 and CD127, while normal melanocytes, only the receptor [53]. The established melanoma cell lines, as well as melanocytes, respond to IL-7 stimulation with increased expression of ICAM-1 [126]. In cell lines harboring BRAF^{V600E} mutation, characterized by the constitutive activation of the MAPK signaling cascade, IL-7 expression is upregulated by MAPK activation and downregulated by the pathway inhibition [55]. In murine melanomas, the tumor growth is stimulated by IL-7, which increases the population of MDSCs

(CD11b+Gr-1+) and IL17-producing γδ27− cells. The effect is mediated by STAT3 and not by STAT5 or PI3K [138]. However, others have shown antitumor activity of IL-7. Primary cultures of melanoma cells transfected with IL-7 have improved sensitivity to immunologic effector cells [227]. In addition, human melanoma cell line, engineered to overexpress IL-7, has displayed substantial tumor growth retardation. Moreover, it has induced CTLs in mixed tumor-lymphocyte cultures and had markedly reduced expression of TGFβ [56]. Recently, Bi et al. [152] demonstrated that IL-7 reduces the number of lung metastases in murine lung melanoma model. The interleukin has stimulated expression of IL-9 and IL-21 in tumor bed, specifically, in the tumor-infiltrating CD4+T cells. In isolated cells, IL-7 induces Th9 differentiation by modifying chromatin structure. At the molecular level, the interleukin increases the expression of histone acetyltransferases and associated proteins, GCN5 and p300, which results in the acetylation of H3K9 at the IL-9 promoter locus. The signaling pathways activated by IL-7/IL-7R include PI3K/Akt, STAT5, and ERK. Of those, the PI3K/Akt/mTOR contributed to the histone acetylation by increasing nuclear accumulation of FoxO1 and its binding with IL-9 promoter and by enhancing translocation of the inhibitory Foxp1 into cytoplasm [152].

2.7.2.11 Breast Cancer

In breast cancer, tumor cells express CD127, but not IL-7, implying a paracrine mode of activation [46, 230]. The interleukin expression in breast cancer tissue is upregulated, indicative of IL-7 overexpression in stromal cells [3]. Also breast tumor interstitial fluid has higher IL-7 concentration than matched normal one [294]. The IL-7 accelerates growth of breast cancer cells in a PI3K- and JAK3-dependent manner, as demonstrated by pathway-specific inhibitors [46]. The interleukin induces VEGF-D expression at mRNA and protein level in human breast cancer cell lines [230]. In clinical samples, the expression of IL-7 and IL-7R is positively correlated with tumor dedifferentiation and cancer advancement, particularly with lymph node involvement,

and negatively with patients' survival. Interestingly, nodal involvement seems to affect the signaling pathway triggered. The N2 cancers have significantly elevated expression of JAK3 and PI3K, as compared to N0 cancers, but decreased expression of JAK1 and STAT5 [3]. Pan et al. [174] and Yang et al. [82] evaluated the effect of IL-7δ5 variant and found it to promote breast cancer cell proliferation and cell cycle progression, by increasing cycling D1 and decreasing p27^{KIP1} expression. Both activities have been dependent on the activation of PI3K/Akt pathway [174]. The PI3K/Akt pathway is also involved in promoting IL-7δ5-mediated EMT. In breast cancer cell lines, the IL-7δ5 decreased E-cadherin and increased N-cadherin expression and induced scattered cell morphology. In xenograft tumors, IL-7δ5 promoted the development of lung metastases and reduced the overall-survival of tumor-bearing mice by 40% [82]. The CD4+ and CD8+ T cells from breast cancer patients has lower expression of CD127 and are markedly less responsive to IL-7 stimulation than T cells from healthy individuals, contributing to the suppressive immunological profile characteristic for patients with solid tumors [295]. In turn, IL-7 in breast cancer patients promotes intratumoral accumulation of CD3^{bright}γδT17 cells, strongly expressing CD127, and enhances their synthesis of protumoral IL-17A, the action antagonized by type I IFN [75].

2.7.2.12 Hepatocellular Carcinoma (HCC)

Kong et al. [54] demonstrated that while CD127 is present on both normal and malignant hepatocytes, IL-7 is expressed exclusively by cancer cells. The authors have shown that HBV infection increases CD127 expression. It involves activation of Notch1 and downstream NF-κB dependent on viral multifunctional nonstructural protein X (HBX). The HBX-activation of IL-7/IL-7R pathway includes phosphorylation of Akt and JNK kinases, but not STAT5, and results in the upregulation of cyclin D and MMP-9, leading to accelerated cell proliferation and migration. Unexpectedly, the IL-7/IL-7R signaling has downregulated vimentin and β-catenin and upregulated the expression of E-cadherin. In clinical samples, CD127 expression has been lower in poorly differentiated tumors [296].

2.0 Conclusions and Perspectives

The functional studies have only started to uncover a protumorigenic activity of IL-7. Little to nothing is known about IL-7/IL-7R signaling in the microenvironment of solid tumors, and our understanding of the pathways involved in hematological malignancies is far from complete. Surprisingly, even the cellular source of the cytokine frequently remains obscure, and there are controversies concerning the ability of particular cell types to express IL-7 and/or its receptor. Nonetheless, the IL-7/IL-7R axis seems to support the development of hematological malignancies by increasing cancer cell viability via mitogenic, trophic and antiapoptotic stimuli, relying mainly on PI3K/AKt and JAK/STAT5 pathways. In solid tumors, apart from prosurvival signals for cancer cells, the IL-7/IL-7R pathway facilitates tumor invasion and metastasis, by enabling extracellular matrix remodeling, epithelial-mesenchymal transition, and lymphangiogenesis, for which it engages also MEK/ERK pathways. Given the unwavering interest in IL-7 application in immunotherapy, based on its immunomodulatory activity, better understanding of its signaling and the cells that are likely to respond should be a research priority.

References

1. Sportès C, Babb RR, Krumlauf MC et al (2010) Phase I study of recombinant human interleukin-7 administration in subjects with refractory malignancy. Clin Cancer Res 16:727–735
2. Bednarz-Misa I, Diakowska D, Krzystek-Korpacka M (2019) Local and systemic IL-7 concentration in gastrointestinal-tract cancers. Medicina (Kaunas) 55:E262
3. Al-Rawi MAA, Rmali K, Watkins G et al (2004) Aberrant expression of interleukin-7 (IL-7) and its signalling complex in human breast cancer. Eur J Cancer 40:494–502
4. Cui L, Fu J, Pang JC-S et al (2012) Overexpression of IL-7 enhances cisplatin resistance in glioma. Cancer Biol Ther 13:496–503
5. Mengus C, Le Magnen C, Trella E et al (2011) Elevated levels of circulating IL-7 and IL-15 in patients with early stage prostate cancer. J Transl Med 9:162
6. Ming J, Zhang Q, Qiu X et al (2009) Interleukin 7/interleukin 7 receptor induce c-Fos/c-Jun-dependent vascular endothelial growth factor-D up-regulation: a mechanism of lymphangiogenesis in lung cancer. Eur J Cancer 45:866–873
7. Suzuki K, Kadota K, Sima CS et al (2013) Clinical impact of immune microenvironment in stage I lung adenocarcinoma: tumor interleukin-12 receptor $\beta2$ (IL-12Rβ2), IL-7R, and stromal FoxP3/CD3 ratio are independent predictors of recurrence. J Clin Oncol 31:490–498
8. Kadota K, Yeh YC, Villena-Vargas J et al (2015) Tumor budding correlates with the protumor immune microenvironment and is an independent prognostic factor for recurrence of stage I lung adenocarcinoma. Chest 148:711–721
9. Kim MJ, Choi SK, Hong SH et al (2018) Oncogenic IL7R is downregulated by histone deacetylase inhibitor in esophageal squamous cell carcinoma via modulation of acetylated FOXO1. Int J Oncol 53:395–403
10. Jiang H, Yang H, Wang H et al (2018) Elevated IL-7Rα is linked to recurrence and poorer survival of gastric adenocarcinoma. Int J Clin Exp Pathol 11:1645–1652
11. Fry TJ, Mackall CL (2005) The many faces of IL-7: from lymphopoiesis to peripheral T cell maintenance. J Immunol 174:6571–6576
12. Jiang Q, Li WQ, Aiello FB et al (2005) Cell biology of IL-7, a key lymphotrophin. Cytokine Growth Factor Rev 16:513–533
13. Kim GY, Hong C, Park J-H (2011) Seeing is believing: illuminating the source of in vivo interleukin-7. Immune Net 11:1–10
14. Lundström W, Fewkes N, Mackall CL (2012) IL-7 in human health and disease. Semin Immunol 24:218–224
15. Oliveira ML, Akkapeddi P, Ribeiro D et al (2019) IL-7R-mediated signaling in T-cell acute lymphoblastic leukemia: an update. Adv Biol Regul 71:88–96
16. Gao J, Zhao L, Wan YY et al (2015) Mechanism of action of IL-7 and its potential applications and limitations in cancer immunotherapy. Int J Mol Sci 16:10267–10280
17. Lin J, Zhu Z, Xiao H et al (2017) The role of IL-7 in immunity and cancer. Anticancer Res 37:963–967
18. Hanahan D, Weinberg RA (2011) Hallmarks of cancer: the next generation. Cell 144:646–674
19. Vudattu NK, Magalhaes I, Hoehn H et al (2009) Expression analysis and functional activity of interleukin-7 splice variants. Genes Immun 10:132–140

20. Park LS, Friend DJ, Schmierer AE et al (1990) Murine interleukin 7 (IL-7) receptor. Characterization on an IL-7-dependent cell line. J Exp Med 171:1073–1089

21. Girard D, Beaulieu AD (1997) Absence of the IL-7 receptor component CDw127 indicates that gamma(c) expression alone is insufficient for IL-7 to modulate human neutrophil responses. Clin Immunol Immunopathol 83:264–271

22. Akashi K, Traver D, Miyamoto T, Weissman IL (2000) A clonogenic common myeloid progenitor that gives rise to all myeloid lineages. Nature 404:193–197

23. Meier D, Bornmann C, Chappaz S et al (2007) Ectopic lymphoid-organ development occurs through interleukin 7-mediated enhanced survival of lymphoid-tissue-inducer cells. Immunity 26:643–654

24. Iolyeva M, Aebischer D, Proulx ST et al (2013) Interleukin-7 is produced by afferent lymphatic vessels and supports lymphatic drainage. Blood 122:2271–2281

25. Onder L, Narang P, Scandella E et al (2012) IL-7-producing stromal cells are critical for lymph node remodeling. Blood 120:4675–4683

26. Rich BE, Campos-Torres J, Tepper RI et al (1993) Cutaneous lymphoproliferation and lymphomas in interleukin 7 transgenic mice. J Exp Med 177:305–316

27. Overell RW, Clark L, Lynch D et al (1991) Interleukin-7 retroviruses transform pre-B cells by an autocrine mechanism not evident in Abelson murine. Mol Cell Biol 11:1590–1597

28. Zamisch M, Moore-Scott B, Su DM et al (2005) Ontogeny and regulation of IL-7-expressing thymic epithelial cells. J Immunol 174:60–67

29. Alves NL, Goff OR-L, Huntington ND et al (2009) Characterization of the thymic IL-7 niche in vivo. Proc Natl Acad Sci U S A 106:1512–1517

30. Scupoli MT, Vinante F, Krampera M (2003) Thymic epithelial cells promote survival of human T-cell acute lymphoblastic leukemia blasts: the role of interleukin-7. Haematologica 88:1229–1237

31. Gray DHD, Tull D, Ueno T et al (2007) A unique thymic fibroblast population revealed by the monoclonal antibody MTS-15. J Immunol 178:4956–4965

32. Cordeiro Gomes A, Hara T, Lim VY et al (2016) Hematopoietic stem cell niches produce lineage-instructive signals to control multipotent progenitor differentiation. Immunity 45:1219–1231

33. Scupoli MT, Perbellini O, Krampera M et al (2007) Interleukin 7 requirement for survival of T-cell acute lymphoblastic leukemia and human thymocytes on bone marrow stroma. Haematologica 92:264–266

34. Zhang X, Tu H, Yang Y et al (2016) High IL-7 levels in the bone marrow microenvironment mediate imatinib resistance and predict disease progression in chronic myeloid leukemia. Int J Hematol 104:358–367

35. Weitzmann MN, Cenci S, Rifas L et al (2000) Interleukin-7 stimulates osteoclast formation by up-regulating the T-cell production of soluble osteoclastogenic cytokines. Blood 96:1873–1878

36. Giuliani N, Colla S, Sala R et al (2002) Human myeloma cells stimulate the receptor activator of nuclear factor-kappa B ligand (RANKL) in T lymphocytes: a potential role in multiple myeloma bone disease. Blood 100:4615–4621

37. Kröncke R, Loppnow H, Flad HD et al (1996) Human follicular dendritic cells and vascular cells produce interleukin-7: a potential role for interleukin-7 in the germinal center reaction. Eur J Immunol 26:2541–2544

38. Takakuwa T, Nomura S, Matsuzuka F et al (2000) Expression of interleukin-7 and its receptor in thyroid lymphoma. Lab Investig 80:1483–1490

39. Link A, Vogt TK, Favre S et al (2007) Fibroblastic reticular cells in lymph nodes regulate the homeostasis of naive T cells. Nat Immunol 8:1255–1265

40. Fujihashi K, Kawabata S, Hiroi T (1996) Interleukin 2 (IL-2) and interleukin 7 (IL-7) reciprocally induce IL-7 and IL-2 receptors on gamma delta T-cell receptor-positive intraepithelial lymphocytes. Proc Natl Acad Sci U S A 93:3613–3618

41. Sorg RV, McLellan AD, Hock BD et al (1998) Human dendritic cells express functional interleukin-7. Immunobiology 198:514–526

42. de Saint-Vis B, Fugier-Vivier I, Massacrier C et al (1998) The cytokine profile expressed by human dendritic cells is dependent on cell subtype and mode of activation. J Immunol 160:1666–1676

43. Mumprecht S, Schürch C, Scherrer S et al (2010) Chronic myelogenous leukemia maintains specific CD8(+) T cells through IL-7 signaling. Eur J Immunol 40:2720–2730

44. Di Carlo E, D'Antuono T, Pompa P et al (2009) The lack of epithelial interleukin-7 and BAFF/BLyS gene expression in prostate cancer as a possible mechanism of tumor escape from immunosurveillance. Clin Cancer Res 15:2979–2987

45. Madrigal-Estebas L, McManus R, Byrne B et al (1997) Human small intestinal epithelial cells secrete interleukin-7 and differentially express two different interleukin-7 mRNA transcripts: implications for extrathymic T-cell differentiation. Hum Immunol 58:83–90

46. Al-Rawi MAA, Rmali K, Mansel RE et al (2004) Interleukin 7 induces the growth of breast cancer cells through a Wortmannin-sensitive pathway. Br J Surg 91:61–68

47. Qu H, Zou Z, Pan Z et al (2016) IL-7/IL-7 receptor axis stimulates prostate cancer cell invasion and migration via AKT/NF-κB pathway. Int Immunopharmacol 40:203–210

48. Heufler C, Topar G, Grasseger A et al (1993) Interleukin 7 is produced by murine and human keratinocytes. J Exp Med 178:1109–1114

49. Long D, Blake S, Song XY et al (2008) Human articular chondrocytes produce IL-7 and respond to IL-7 with increased production of matrix metalloproteinase-13. Arthritis Res Ther 10:R23

50. Moors M, Vudattu NK, Abel J et al (2010) Interleukin-7 (IL-7) and IL-7 splice variants affect differentiation of human neural progenitor cells. Genes Immun 11:11–20

51. Szot PA, Franklin DP, Figlewicz TP et al (2017) Multiple lipopolysaccharide (LPS) injections alters interleukin 6 (IL 6), IL 7, IL 10 and IL 6 and IL 7 receptor mRNA in CNS and spleen. Neuroscience 355:9–21

52. Jana M, Mondal S, Jana A et al (2014) Interleukin-12 (IL-12), but not IL-23, induces the expression of IL-7 in microglia and macrophages: implications for multiple sclerosis. Immunology 141:549–563

53. Mattei S, Colombo MP, Melani C et al (1994) Expression of cytokine/growth factors and their receptors in human melanoma and melanocytes. Int J Cancer 56:853–857

54. Kong F, Hu W, Zhou K et al (2016) Hepatitis B virus X protein promotes interleukin-7 receptor expression via NF-κB and Notch1 pathway to facilitate proliferation and migration of hepatitis B virus-related hepatoma cells. J Exp Clin Cancer Res 35:172

55. Whipple CA, Boni A, Fisher JL et al (2016) The mitogen-activated protein kinase pathway plays a critical role in regulating immunological properties of BRAF mutant cutaneous melanoma cells. Melanoma Res 26:223–235

56. Miller AR, McBride WH, Dubinett SM et al (1993) Transduction of human melanoma cell lines with the human interleukin-7 gene using retroviral-mediated gene transfer: comparison of immunologic properties with interleukin-2. Blood 82:3686–3694

57. Cattaruzza L, Gloghini A, Olivo K et al (2009) Functional coexpression of interleukin (IL)-7 and its receptor (IL-7R) on Hodgkin and Reed-Sternberg cells: involvement of IL-7 in tumor cell growth and microenvironmental interactions of Hodgkin's lymphoma. Int J Cancer 125:1092–1101

58. Jian M, Yunjia Z, Zhiying D et al (2019) Interleukin 7 receptor activates PI3K/Akt/mTOR signaling pathway via downregulation of Beclin-1 in lung cancer. Mol Carcinog 58:358–365

59. Takuechi Y, Yamanouchi H, Yue Q et al (1998) Epithelial component of lymphoid stroma-rich Warthin's tumour expresses interleukin (IL)-7. Histopathology 32:383–384

60. Hahtola S, Tuomela S, Elo L et al (2006) Th1 response and cytotoxicity genes are down-regulated in cutaneous T-cell lymphoma. Clin Cancer Res 12:4812–4821

61. Iwata M, Graf L, Awaya N et al (2002) Functional interleukin-7 receptors (IL-7Rs) are expressed by marrow stromal cells: binding of IL-7 increases levels of IL-6 mRNA and secreted protein. Blood 100:1318–1325

62. Pillai M, Torok-Storb B, Iwata M (2004) Expression and function of IL-7 receptors in marrow stromal cells. Leuk Lymphoma 45:2403–2408

63. Musso T, Johnston JA, Linnekin D et al (1995) Regulation of JAK3 expression in human mono-cytes: phosphorylation in response to interleukins 2, 4, and 7. J Exp Med 181:1425–1431

64. Watanabe M, Ueno Y, Yajima T et al (1995) Interleukin-7 is produced by human intestinal epithelial-cells and regulates the proliferation of intestinal mucosal lymphocytes. J Clin Invest 95:2945–2953

65. Mazzucchelli R, Durum SK (2007) Interleukin-7 receptor expression: intelligent design. Nat Rev Immunol 7:144–154

66. Nishida C, Kusubata K, Tashiro Y et al (2012) MT1-MMP plays a critical role in hematopoiesis by regulating HIF-mediated chemokine/cytokine gene transcription within niche cells. Blood 119:5405–5416

67. Sawa Y, Arima Y, Ogura H et al (2009) Hepatic interleukin-7 expression regulates T cell responses. Immunity 30:447–457

68. Oshima S, Nakamura T, Namiki S et al (2004) Interferon regulatory factor 1 (IRF-1) and IRF-2 distinctively up-regulate gene expression and production of interleukin-7 in human intestinal epithelial cells. Mol Cell Biol 24:6298–6310

69. Harada S, Yamamura M, Okamoto H et al (1999) Production of interleukin-7 and interleukin-15 by fibroblast-like synoviocytes from patients with rheumatoid arthritis. Arthritis Rheum 42:1508–1516

70. Saha A, Blando J, Silver E et al (2014) 6-Shogaol from dried ginger inhibits growth of prostate cancer cells both *in vitro* and *in vivo* through inhibition of STAT3 and NF-kB signaling. Cancer Prev Res (Phila) 7:627–638

71. Trinder P, Seitzer U, Gerdes J et al (1999) Constitutive and IFN-y regulated expression of IL-7 and IL-15 in human renal cell cancer. Int J Oncol 14:23–31

72. Cai YJ, Wang WS, Yang Y et al (2013) Up-regulation of intestinal epithelial cell derived IL-7 expression by keratinocyte growth factor through STAT1/IRF-1, IRF-2 pathway. PLoS One 8:e58647

73. Tang J, Nuccie BL, Ritterman I et al (1997) TGF-beta down-regulates stromal IL-7 secretion and inhibits proliferation of human B cell precursors. J Immunol 159:117–125

74. Nye MD, Almada LL, Fernandez-Barrena MG et al (2014) The transcription factor GLI1 interacts with SMAD proteins to modulate transforming growth factor β-induced gene expression in a p300/CREB-binding protein-associated factor (PCAF)-dependent manner. J Biol Chem 289:15495–15506

75. Patin EC, Soulard D, Fleury et al (2018) Type I IFN receptor signaling controls IL7-dependent accumulation and activity of protumoral IL17A-producing γδT cells in breast cancer. Cancer Res 78:195–204

76. Korte A, Köchling J, Badiali L et al (2000) Expression analysis and characterization of alternatively spliced transcripts of human IL-7Ralpha chain encoding two truncated receptor proteins in relapsed childhood all. Cytokine 12:1597–1608

77. Michaud A, Dardari R, Charrier E et al (2010) IL-7 enhances survival of human CD56bright NK cells. J Immunother 33:382–390

78. Maude SL, Dolai S, Delgado-Martin C et al (2015) Efficacy of JAK/STAT pathway inhibition in murine xenograft models of early T-cell precursor (ETP) acute lymphoblastic leukemia. Blood 125:1759–1767

79. Cosenza L, Gorgun G, Urbano A et al (2002) Interleukin-7 receptor expression and activation in nonhaematopoietic neoplastic cell lines. Cell Signal 14:317–325

80. Maeurer MJ, Walter W, Martin D et al (1997) Interleukin-7 (IL-7) in colorectal cancer: IL-7 is produced by tissues from colorectal cancer and promotes preferential expansion of tumour infiltrating lymphocytes. Scand J Immunol 45:182–192

81. Jian M, Qingfu Z, Yanduo J (2015) Anti-lymphangiogenesis effects of a specific anti-interleukin 7 receptor antibody in lung cancer model in vivo. Mol Carcinog 54:148–155

82. Yang J, Zeng Z, Peng Y et al (2014) IL-7 splicing variant IL-7delta5 induces EMT and metastasis of human breast cancer cell lines MCF-7 and BT-20 through activation of PI3K/Akt pathway. Histochem Cell Biol 142:401–410

83. Laouar Y, Crispe IN, Flavell RA (2004) Overexpression of IL-7Rα provides a competitive advantage during early T-cell development. Blood 103:1985–1994

84. Cramer SD, Hixon JA, Andrews C et al (2018) Mutant IL-7Rα and mutant NRas are sufficient to induce murine T cell acute lymphoblastic leukemia. Leukemia 32:1795–1882

85. Henriques CM, Rino J, Nibbs RJ et al (2010) IL-7 induces rapid clathrin-mediated internalization and JAK3-dependent degradation of IL-7Ralpha in T cells. Blood 115:3269–3277

86. Sasson SC, Smith S, Seddiki N et al (2010) IL-7 receptor is expressed on adult pre-B-cell acute lymphoblastic leukemia and other B-cell derived neoplasms and correlates with expression of proliferation and survival markers. Cytokine 50:58–68

87. Mazzucchelli RI, Riva A, Durum SK (2012) The human IL-7 receptor gene: deletions, polymorphisms and mutations. Semin Immunol 24:225–230

88. Vitiello GAF, Losi Guembarovski R, Amarante MK et al (2018) Interleukin 7 receptor alpha Thr244Ile genetic polymorphism is associated with susceptibility and prognostic markers in breast cancer subgroups. Cytokine 103:121–126

89. Zenatti PP, Ribeiro D, Li W et al (2011) Oncogenic IL7R gain-of-function mutations in childhood T-cell acute lymphoblastic leukemia. Nat Genet 43:932–939

90. Kim MS, Chung NG, Kim MS et al (2013) Somatic mutation of IL7R exon 6 in acute leukemias and solid cancers. Hum Pathol 44:551–555

91. Goodwin RG, Friend D, Ziegler SF et al (1990) Cloning of the human and murine interleukin-7 receptors: demonstration of a soluble form and homology to a new receptor superfamily. Cell 60:941–951

92. Crawley AM, Vranjkovic A, Young C et al (2010) Interleukin-4 downregulates CD127 expression and activity on human thymocytes and mature CD8+ T cells. Eur J Immunol 40:1396–1407

93. Cui H, Xie N, Tan Z et al (2014) The human long noncoding RNA lnc-IL7R regulates the inflammatory response. Eur J Immunol 44:2085–2095

94. Ye Z, Xu J, Li S et al (2017) Lnc-IL7R promotes the growth of fibroblast-like synoviocytes through interaction with enhancer of zeste homolog 2 in rheumatoid arthritis. Mol Med Rep 15:1412–1418

95. Ding L, Ren J, Zhang D et al (2017) The TLR3 agonist inhibit drug efflux and sequentially consolidates low-dose cisplatin-based chemoimmunotherapy while reducing side effects. Mol Cancer Ther 16:1068–1079

96. Fan Y, Yan N, Juanjuan H et al (2018) Up-regulation of inflammation-related LncRNA-IL7R predicts poor clinical outcome in patients with cervical cancer. Biosci Rep 38:BSR20180483

97. Zhang F, Liang X, Pu D et al (2012) Biophysical characterization of glycosaminoglycan-IL-7 interactions using SPR. Biochimie 94:242–249

98. Clarke D, Katoh O, Gibbs RV et al (1995) Interaction of interleukin 7 (IL-7) with glycosaminoglycans and its biological relevance. Cytokine 7:325–330

99. Borghesi LA, Yamashita Y, Kincade PW (1999) Heparan sulfate proteoglycans mediate interleukin-7-dependent B lymphopoiesis. Blood 93:140–148

100. Fu R, Gao S, Peng F et al (2014) Relationship between abnormal osteoblasts and cellular immunity in multiple myeloma. Cancer Cell Int 14:62

101. Kupsa T, Vasatova M, Karesova I, Zak P, Horacek JM (2014) Baseline serum levels of multiple cytokines and adhesion molecules in patients with acute myeloid leukemia: results of a pivotal trial. Exp Oncol 36:252–257

102. Chopra V, Dinh TV, Hannigan EV (1998) Circulating serum levels of cytokines and angiogenic factors in patients with cervical cancer. Cancer Investig 16:152–159

103. Mojtahedi Z, Khademi B, Erfani N et al (2014) Serum levels of interleukin-7 and interleukin-8 in head and neck squamous cell carcinoma. Indian J Cancer 51:227–230

104. Komura T, Sakai Y, Harada K et al (2015) Inflammatory features of pancreatic cancer highlighted by monocytes/macrophages and CD4+ T cells with clinical impact. Cancer Sci 106:672–686

105. Lambeck AJ, Crijns AP, Leffers N et al (2007) Serum cytokine profiling as a diagnostic and prognostic tool in ovarian cancer: a potential role for interleukin 7. Clin Cancer Res 13:2385–2391

106. Provatopoulou X, Georgiadou D, Sergentanis TN et al (2014) Interleukins as markers of inflammation in malignant and benign thyroid disease. Inflamm Res 63:667–674

107. Krzystek-Korpacka M, Zawadzki M, Neubauer K et al (2017) Elevated systemic interleukin-7 in patients with colorectal cancer and individuals at high risk of cancer: association with lymph node involvement and tumor location in the right colon. Cancer Immunol Immunother 66:171–179

108. Janusz L, Filfiniec Pérez J, Visse F et al (2016) Preoperative systemic levels of VEGFA, IL-7, IL-17A, and TNF-β delineate two distinct groups of children with brain tumors. Pediatr Blood Cancer 63:2112–2122

109. Shen Q, Polom K, Williams C et al (2019) A targeted proteomics approach reveals a serum protein signature as diagnostic biomarker for resectable gastric cancer. EBioMedicine 44:322–333

110. Krzystek-Korpacka M, Zawadzki M, Kapturkiewicz B et al (2018) Subsite heterogeneity in the profiles of circulating cytokines in colorectal cancer. Cytokine 110:435–441

111. Bordbar E, Malekzadeh M, Ardekani MT, Doroudchi M, Ghaderi A (2012) Serum levels of G-CSF and IL-7 in Iranian breast cancer patients. Asian Pac J Cancer Prev 13:5307–5312

112. Roato I, Brunetti G, Gorassini E et al (2006) IL-7 up-regulates TNF-α-dependent osteoclastogenesis in patients affected by solid tumor. PLoS One 1:e124

113. Shiels MS, Pfeiffer RM, Hildesheim A et al (2013) Circulating inflammation markers and prospective risk for lung cancer. J Natl Cancer Inst 105:1871–1880

114. Comstock SS, Xu D, Hortos K et al (2016) Association of serum cytokines with colorectal polyp number and type in adult males. Eur J Cancer Prev 25:173–181

115. Bussu F, Graziani C, Gallus R et al (2018) IFN-γ and other serum cytokines in head and neck squamous cell carcinomas. Acta Otorhinolaryngol Ital 38:94–102

116. Lv M, Xiaoping X, Cai H et al (2011) Cytokines as prognstic tool in breast carcinoma. Front Biosci (Landmark Ed) 16:2515–2526

117. Chen ZY, He WZ, Peng LX et al (2015) A prognostic classifier consisting of 17 circulating cytokines is a novel predictor of overall survival for metastatic colorectal cancer patients. Int J Cancer 136:584–592

118. Lin Y, Zhang G, Zhang J et al (2013) A panel of four cytokines predicts the prognosis of patients with malignant gliomas. J Neuro-Oncol 114:199–208

119. Cheng Y, Zheng S, Pan CT et al (2017) Analysis of aqueous humor concentrations of cytokines in retinoblastoma. PLoS One 12:e0177337

120. Ortiz Franyuti D, Mitsi M, Vogel V (2017) Mechanical stretching of fibronectin fibers upregulates binding of interleukin-7. Nano Lett 18:15–25

121. Oritani K, Kinkade PW (1996) Identification of stromal cell products that interact with pre-B cells. J Cell Biol 134:771–782

122. Lai L, Goldschneider I (2001) Cutting edge: identification of a hybrid cytokine consisting of IL-7 and the beta-chain of the hepatocyte growth factor/scatter factor. J Immunol 167:3550–3554

123. Ariel A, Hershkoviz R, Cahdon L et al (1997) Induction of T cell adhesion to extracellular matrix or endothelial cell ligands by soluble or matrix-bound inte rleukin-7. Eur J Immunol 27:2562–2570

124. Limbro R, Varrena J, Arthor J et al (2012) IL-7 induces expression and activation of integrin A4β7 promoting naive T-cell homing to the intestinal mucosa. Blood 120:2610–2619

125. Xie X, Ye D, Chen H et al (2004) Effect of interleukin-7 gene transfection into ovarian carcinoma cell line SKOV3 in vitro and *in vivo*. Gynecol Oncol 92:578–585

126. Kirnbauer B, Charvat B, Shauer E et al (1992) Modulation of intercellular adhesion molacule-1 expression on human melanocytes and melanoma cells: evidence for a regulator role of IL-6, IL-7, TNF-beta and UVB light. J Invest Dermatol 98:320–326

127. Zhang L, Keane MP, Zhu LX et al (2004) Interleukin-7 and transforming growth factor-beta play counter-regulatory roles in protein kinase C-delta-dependent control of fibroblast collagen synthesis in pulmonary fibrosis. J Biol Chem 279:28315–28319

128. Yamanaka O, Saika S, Ikeda K et al (2006) Interleukin-7 modulates extracellular matrix production and TGF-beta signaling in cultured human subconjunctival fibroblasts. Curr Eye Res 31:491–499

129. Yammani RR, Long D, Loeser RF (2009) Interleukin-7 stimulates secretion of S100A4 by activating the JAK/STAT signaling pathway in human articular chondrocytes. Arthritis Rheum 60:792–800

130. Park SL, Lee EJ, Kim WJ et al (2014) p27KIP1 is involved in ERK1/2-mediated MMP-9 expression via the activation of NF-kappaB binding in the IL-7-induced migration and invasion of 5637 cells. Int J Oncol 44:1349–1356

131. Shochat C, Tal N, Bandapalli OR et al (2011) Gain-of-function mutations in interleukin-7 receptor-α (IL7R) in childhood acute lymphoblastic leukemias. J Exp Med 208:901–908

132. Shochat C, Tal N, Gryshkova V et al (2014) Novel activating mutations lacking cysteine in type I cytokine receptors in acute lymphoblastic leukemia. Blood 124:106–110

133. Degryse S, de Bock CE, Demeyer S et al (2018) Mutant JAK3 phosphoproteomic profiling predicts synergism between JAK3 inhibitors and MEK/BCL2 inhibitors for the treatment of T-cell acute lymphoblastic leukemia. Leukemia 32:788–800

134. Huang M, Sharma S, Zhu LX et al (2002) IL-7 inhibits fibroblast TGF-beta production and signaling in pulmonary fibrosis. J Clin Invest 109:931–937

135. van der Plas DC, Smiers F, Pouwels K et al (1996) Interleukin-7 signaling in human B cell precursor acute lymphoblastic leukemia cells and murine BAF3 cells involves activation of STAT1 and STAT5 mediated via the interleukin-7 receptor alpha chain. Leukemia 10:1317–1325

136. Pike KA, Hatzihristidis T, Bussières-Marmen S et al (2017) TC-PTP regulates the IL-7 transcriptional response during murine early T cell development. Sci Rep 7:13275

137. Chou WC, Levy DE, Lee CK (2006) STAT3 positively regulates an early step in B-cell development. Blood 108:3005–3011

138. Li J, Liu J, Mao X et al (2014) IL-7 receptor blockade inhibits IL-17-producing γδ cells and suppresses melanoma development. Inflammation 37:1444–1452

139. Corfe SA, Paige CJ (2012) The many roles of IL-7 in B cell development; mediator of survival, proliferation and differentiation. Semin Immunol 24:198–208

140. Ribeiro D, Melão A, van Boxtel R et al (2018) STAT5 is essential for IL-7–mediated viability, growth, and proliferation of T-cell acute lymphoblastic leukemia cells. Blood Adv 2:2199–2213

141. Qin JZ, Zhang CL, Kamarashev J et al (2001) Interleukin-7 and interleukin-15 regulate the expression of the bcl-2 and c-myb genes in cutaneous T-cell lymphoma cells. Blood 98:2778–2783

142. Melão A, Spit M, Cardoso BA et al (2016) Optimal interleukin-7 receptor-mediated signaling, cell cycle progression and viability of T-cell acute lymphoblastic leukemia cells rely on casein kinase 2 activity. Haematologica 101:1368–1379

143. Zeng H, Yu M, Tan H et al (2018) Discrete roles and bifurcation of PTEN signaling and mTORC1-mediated anabolic metabolism underlie IL-7–driven B lymphopoiesis. Sci Adv 4:eaar5701

144. Nagel S, Pommerenke C, Meyer C et al (2016) Deregulation of polycomb repressor complex 1 modifier AUTS2 in T-cell leukemia. Oncotarget 7:45398–45413

145. Drake A, Kaur M, Iliopoulou BP et al (2016) Interleukins 7 and 15 maintain human T cell proliferative capacity through STAT5 signaling. PLoS One 11:e0166280

146. Pallard C, Stegmann AP, van Kleffens T et al (1999) Distinct rôles of the phosphatidylinositol 3-kinase and STAT5 pathways in IL-7-mediated development of human thymocyte precursors. Immunity 10:525–535

147. Crawley AM, Vranjkovic A, Faller E et al (2013) Jak/STAT and PI3K signaling pathways have both common and distinct roles in IL-7-mediated activities in human CD8+ T cells. J Leukoc Biol 95:117–127

148. Kim HK, Waickman AT, Castro E et al (2016) Distinct IL-7 signaling in recent thymic emigrants versus mature naïve T cells controls T-cell homeostasis. Eur J Immunol 46:1669–1680

149. Zhang X, Tu H, Yang Y et al (2019) Bone marrow-derived mesenchymal stromal cells promote resistance to tyrosine kinase inhibitors in chronic myeloid leukemia via the IL-7/JAK1/STAT5 pathway. J Biol Chem 294:12167–12179

150. Delgado-Martin C, Meyer LK, Huang BJ et al (2017) JAK/STAT pathway inhibition overcomes IL7-induced glucocorticoid resistance in a subset of human T-cell acute lymphoblastic leukemias. Leukemia 31:2568–2576

151. Ding ZC, Liu C, Cao Y et al (2016) IL-7 signaling imparts polyfunctionality and stemness potential to CD4(+) T cells. Onco Targets Ther 5:e1171445

152. Bi E, Ma X, Lu Y et al (2017) Foxo1 and Foxp1 play opposing roles in regulating the differentiation and antitumor activity of TH9 cells programmed by IL-7. Sci Signal 10(500):eaak9741

153. Ribeiro D, Melão A, van Boxtel R et al (2018) IL-7 activates a STAT5/PIM1 axis to promote T-cell acute lymphoblastic leukemia proliferation and viability in a bcl-2-independent manner. Blood 132:914

154. Seol MA, Kim JH, Oh K et al (2019) Interleukin-7 contributes to the invasiveness of prostate cancer cells by promoting epithelial-mesenchymal transition. Sci Rep 9:6917

155. Wofford JA, Wieman HL, Jacobs SR et al (2008) IL-7 promotes GLUT1 trafficking and glucose uptake via STAT5-mediated activation of Akt to support T-cell survival. Blood 111:2101–2111

156. Goetz CA, Harmon IR, O'Neil JJ et al (2004) STAT5 activation underlies IL7 receptor-dependent B cell development. J Immunol 172:4770–4778

157. Zhang X, Song M, Kundu JK et al (2018) PIM kinase as an executional target in cancer. J Cancer Prev 23:109–116

158. Hiasa M, Teramachi J, Oda A et al (2015) Pim-2 kinase is an important target of treatment for tumor progression and bone loss in myeloma. Leukemia 29:207–217

159. Patra AK, Avots A, Zahedi RP et al (2013) An alternative NFAT-activation pathway mediated by IL-7 is critical for early thymocyte development. Nat Immunol 14:127–135

160. Pan M-G, Xiong Y, Chen F (2013) NFAT gene family in inflammation and cancer. Curr Mol Med 13:543–554

161. Oestreich KJ, Yoon H, Ahmed R et al (2008) NFATc1 regulates PD-1 expression upon T cell activation. J Immunol 181:4832–4839

162. Tripathi MK, Deane NG, Zhu J et al (2014) NFAT transcriptional activity is associated with metastatic capacity in colon cancer. Cancer Res 74:6947–6957

163. Oikawa T, Nakamura A, Onishi N et al (2013) Acquired expression of NFATc1 downregulates E-cadherin and promotes cancer cell invasion. Cancer Res 73:5100–5109

164. Gast CE, Silk AD, Zarour L et al (2018) Cell fusion potentiates tumor heterogeneity and reveals circulating hybrid cells that correlate with stage and survival. Sci Adv 4:eaat7828

165. Sharfe N, Dadi HK, Roifman CM (1995) JAK3 protein tyrosine kinase mediates interleukin-7-induced activation of phosphatidylinositol-3′ kinase. Blood 86:2077–2085

166. Aiello FB, Guszczynski T, Li W et al (2018) IL-7-induced phosphorylation of the adaptor Crk-like and other targets. Cell Signal 47:131–141

167. Ghazawi FM, Faller EM, Sugden SM et al (2016) IL-7 downregulates IL-7Ra expression in human CD8 T cells by two independent mechanisms. Immunol Cell Biol 91:149–158

168. Cornish AL, Chong MM, Davey GM et al (2003) Suppressor of cytokine signaling-1 regulates signaling in response to interleukin 2 and other γc-dependent cytokines in peripheral T cells. J Biol Chem 278:22755–22761

169. Trengove MC, Ward AC (2013) SOCS proteins in development and disease. Am J Clin Exp Immunol 2:1–29

170. Canté-Barrett K, Spijkers-Hagelstein JA, Buijs-Gladdines JG et al (2016) MEK and PI3K-AKT inhibitors synergistically block activated IL7 receptor signaling in T-cell acute lymphoblastic leukemia. Leukemia 30:1832–1843

171. Barata JT, Cardoso AA, Boussiotis VA (2005) Interleukin-7 in T-cell acute lymphoblastic leukemia: an extrinsic factor supporting leukemogenesis? Leuk Lymphoma 46:483–495

172. Batista A, Barata JT, Raderschall E et al (2011) Targeting of active mTOR inhibits primary leukemia T cells and synergizes with cytotoxic drugs and signaling inhibitors. Exp Hematol 39:457–472

173. Li Y, Buijs-Gladdines JG, Canté-Barrett K et al (2016) IL-7 receptor mutations and steroid resistance in pediatric T cell acute lymphoblastic leukemia: a genome sequencing study. PLoS Med 13:e1002200

174. Pan D, Liu B, Jin X et al (2012) IL-7 splicing variant IL-7delta5 induces human breast cancer cell proliferation via activation of PI3K/Akt pathway. Biochem Biophys Res Commun 422:727–731

175. Al-Rawi MAA, Watkins G, Mansel RE et al (2005) The effects of interleukin-7 on the lymphangiogenic properties of human endothelial cells. Int J Oncol 27:721–730

176. Ksionda O, Melton AA, Bache J et al (2016) RasGRP1 overexpression in T-ALL increases basal nucleotide exchange on Ras rendering the Ras/PI3K/Akt pathway responsive to protumorigenic cytokines. Oncogene 35:3658–3668

177. Crawley JB, Rawlinson L, Lali FV et al (1997) T cell proliferation in response to interleukins 2 and 7 requires p38MAP kinase activation. J Biol Chem 272:15023–15027

178. Crawley JB, Willcocks J, Foxwell BM (1996) Interleukin-7 induces T cell proliferation in the absence of ErWMAP kinase activity. Eur J Immunol 26:2717–2723

179. Fleming HE, Paige CJ (2001) Pre-B cell receptor signaling mediates selective response to IL-7 at the pro-B to pre-B cell transition via an ERK/MAP kinase-dependent pathway. Immunity 15:521–531

180. Barata JT, Silva A, Brandao JG et al (2004) Activation of PI3K is indispensable for interleukin 7–mediated viability, proliferation, glucose use, and growth of T cell acute lymphoblastic leukemia cells. J Exp Med 200:659–669

181. Kariminia A, Ivison SM, Leung VM et al (2017) Y-box-binding protein 1 contributes to IL-7-mediated survival signaling in B-cell precursor acute lymphoblastic leukemia. Oncol Lett 13:497–505

182. Martelli AM, Tabellini G, Ricci F et al (2012) PI3K/AKT/mTORC1 and MEK/ERK signaling in T-cell acute lymphoblastic leukemia: new options for targeted therapy. Adv Biol Regul 52:214–227

183. Page TH, Lali FL, Foxwell BMJ (1995) Interleukin-7 activates p56[lck] and p59[fyn], two tyrosine kinases associated with the p90 interleukin-7 receptor in primary human T cells. Eur J Immunol 25:2956–2960

184. Abraham KM, Levin SD, Marth JD et al (1991) Thymic tumorigenesis induced by overexpression of p56[lck]. Proc Natl Acad Sci U S A 88:3977–3981

185. Tycko B, Smith SD, Sklar J (1991) Chromosomal translocations joining LCK and TCRB loci in human T cell leukemia. J Exp Med 174:867–873

186. Seckinger P, Fougereau M (1994) Activation of src family kinases in human pre-B cells by IL-7. J Immunol 153:97–109

187. Venkitaraman AR, Cowling RJ (1992) Interleukin 7 receptor functions by recruiting the tyrosine kinase p59fyn through a segment of its cytoplasmic tail. Proc Natl Acad Sci U S A 89:12083–12087

188. Li Y, Chen W, Ren J et al (2003) DF3/MUC1 signaling in multiple myeloma cells is regulated by interleukin-7. Cancer Biol Ther 2:187

189. Stroopinsky D, Kufe D, Avigan D (2016) MUC1 in hematological malignancies. Leuk Lymphoma 57:2489–2498

190. Shang S, Fang Hua F, Hu Z-W (2017) The regulation of β-catenin activity and function in cancer: therapeutic opportunities. Oncotarget 8:33972–33989

191. Benbernou N, Muegge K, Durum SK (2000) Interleukin (IL)-7 induces rapid activation of Pyk2, which is bound to Janus kinase 1 and IL-7Ra. J Biol Chem 275:7060–7065

192. Pellegrini M, Bouillet P, Robati M (2004) Loss of Bim increases T cell production and function in interleukin 7 receptor–deficient mice. J Exp Med 200:1189–1195

193. Rathmell JC, Farkash EA, Gao W et al (2001) IL-7 enhances the survival and maintains the size of naive T cells. J Immunol 167:6869–6876

194. Khaled AR, Kim K, Hofmeister R et al (1999) Withdrawal of IL-7 induces Bax translocation from cytosol to mitochondria through a rise in intracellular pH. Proc Natl Acad Sci U S A 96:14476–14481

195. Kim K, Lee CK, Sayers TJ et al (1998) The trophic action of IL-7 on pro-T cells: inhibition of apoptosis of pro-T1, -T2, and -T3 cells correlates with Bcl-2 and bax levels and is independent of fas and p53 pathways. J Immunol 160:5735–5741

196. Karawajew L, Ruppert V, Wuchter C et al (2000) Inhibition of in vitro spontaneous apoptosis by IL-7 correlates with Bcl-2 up-regulation, cortical/mature immunophenotype, and better early cytoreduction of childhood T-cell acute lymphoblastic leukemia. Blood 96:297–306

197. Barata JT, Cardoso AA, Nadler LM et al (2001) Interleukin-7 promotes survival and cell cycle progression of T-cell acute lymphoblastic leukemia cells by down-regulating the cyclin-dependent kinase inhibitor p27(kip1). Blood 98:1524–1531

198. Silva A, Laranjeira AB, Martins LR et al (2011) IL-7 contributes to the progression of human T-cell acute lymphoblastic leukemias. Cancer Res 71:4780–4789

199. Opferman JT, Letai A, Beard C et al (2003) Development and maintenance of B and T lymphocytes requires antiapoptotic MCL-1. Nature 426:671–676

200. Li WQ, Jiang Q, Khaled AR et al (2004) Interleukin-7 inactivates the pro-apoptotic protein bad promoting T cell survival. J Biol Chem 279:29160–29166

201. Li WQ, Guszczynski T, Hixon JA et al (2010) Interleukin-7 regulates Bim proapoptotic activity in peripheral T-cell survival. Mol Cell Biol 30:590–600

202. Liu ZH, Wang MH, Ren HJ et al (2014) Interleukin 7 signaling prevents apoptosis by regulating Bcl-2 and Bax *via* the p53 pathway in human non-small cell lung cancer cells. Int J Clin Exp Pathol 7:870–881

203. Andersson A, Yang SC, Huang M et al (2009) IL-7 promotes CXCR3 ligand-dependent T cell antitumor reactivity in lung cancer. J Immunol 182:6951–6918

204. Wuchter C, Ruppert V, Schrappe M et al (2002) In vitro susceptibility to dexamethasone- and doxorubicin-induced apoptotic cell death in context of maturation stage, responsiveness to interleukin 7, and early cytoreduction in vivo in childhood T-cell acute lymphoblastic leukemia. Blood 2002(99):4109–4115

205. Kimura MY, Pobezinsky LA, Guinter TI et al (2013) IL-7 signaling must be intermittent, not continuous, during CD8 T cell homeostasis to promote cell survival instead of cell death. Nat Immunol 14:143–151

206. Lum JJ, Schnepple DJ, Nie Z et al (2004) Differential effects of interleukin-7 and interleukin-15 on NK cell anti-human immunodeficiency virus activity. J Virol 78:6033–6042

207. Qin JZ, Dummer R, Burg G et al (1999) Constitutive and interleukin-7/interleukin-15 stimulated DNA binding of Myc, Jun, and novel Myc-like proteins in cutaneous T-cell lymphoma cells. Blood 93:260–267

208. Li WQ, Jiang Q, Aleem E et al (2006) IL-7 promotes T cell proliferation through destabilization of p27Kip1. J Exp Med 203:573–582

209. Digel W, Schmid M, Heil G et al (1991) Human interleukin-7 induces proliferation of neoplastic cells from chronic lymphocytic leukemia and acute leukemias. Blood 78:753–759

210. Ming J, Jiang G, Zhang Q et al (2012) Interleukin-7 up-regulates cyclin D1 via activator protein-1 to promote proliferation of cell in lung cancer. Cancer Immunol Immunother 61:79–88

211. Brown VI, Fang J, Alcorn K et al (2003) Rapamycin is active against B-precursor leukemia in vitro and in vivo, an effect that is modulated by IL-7-mediated signaling. Proc Natl Acad Sci U S A 100:15113–15118

212. Akasaka T, Tsuji K, Kawahira H et al (1997) The role of mel-18, a mammalian Polycomb group gene, during IL-7-dependent proliferation of lymphocyte precursors. Immunity 7:135–146

213. van der Lugt NMT, Domen J, Linders K et al (1994) Posterior transformation, neurological abnormalities, and severe hematopoietic defects in mice with a targeted deletion of the bmi-1 proto-oncogene. Genes Dev 8:757–769

214. Hartzell C, Ksionda O, Lemmens E et al (2013) Dysregulated RasGRP1 responds to cytokine receptor input in T cell leukemogenesis. Sci Signal 6:ra21

215. Shaw LM (2011) The insulin receptor substrate (IRS) proteins at the intersection of metabolism and cancer. Cell Cycle 10:1750–1756

216. Bauer DE, Harris MH, Plas DR et al (2004) Cytokine stimulation of aerobic glycolysis in hematopoietic cells exceeds proliferative demand. FASEB J 18:1303–1305

217. Silva A, Gírio A, Cebola I et al (2011) Intracellular reactive oxygen species are essential for PI3K/Akt/mTOR-dependent IL-7-mediated viability of T-cell acute lymphoblastic leukemia cells. Leukemia 25:960–967

218. Csibi A, Blenis J (2011) Appetite for destruction: the inhibition of glycolysis as a therapy for tuberous sclerosis complex-related tumors. BMC Biol 9:69

219. Pearson C, Silva A, Seddon B (2012) Exogenous amino acids are essential for interleukin-7 induced CD8 T cell growth. PLoS One 7:e33998

220. Cui G, Staron MM, Gray SM et al (2015) IL-7-induced glycerol transport and TAG synthesis promotes memory CD8+ T cell longevity. Cell 161:750–761

221. Dongre A, Weinberg RA (2019) New insights into the mechanisms of epithelial-mesenchymal transition and implications for cancer. Nat Rev Mol Cell Biol 20:69–84

222. Shilo A, Siegfried Z, Karni R (2015) The role of splicing factors in deregulation of alternative splicing during oncogenesis and tumor progression. Mol Cell Oncol 2:e970955

223. Fei F, Qu J, Zhang M et al (2017) S100A4 in cancer progression and metastasis: a systematic review. Oncotarget 8:73219–73239

224. Cen B, Xiong Y, Song JH et al (2014) The Pim-1 protein kinase is an important regulator of MET receptor tyrosine kinase levels and signaling. Mol Cell Biol 34:2517–2532

225. Goossens S, Radaelli E, Blanchet O et al (2015) ZEB2 drives immature T-cell lympho-blastic leukaemia development via enhanced tumour-initiating potential and IL-7 receptor sig-nalling. Nat Commun 6:5794

226. Dubinett SM, Huang M, Dhanani S et al (1995) Down-regulation of murine fibrosarcoma transforming growth factor-β1 expression by interleukin 7. J Natl Cancer Inst 87:593–597

227. Finke S, Trojanek B, Moller P et al (1997) Increase of sensitivity of primary human melanoma cells

transfected with interleukin-7 gene to autologous and allogenic immunologic effect cells. Cancer Gene Ther 4:260–268

228. Rutto KV, Lyamina IV, Kudryavtsev IV et al (2016) Regulation of vascular endothelial growth factor (VEGF) production by mouse thymic epithelial cell lines. Toxicology 50:106–110

229. Ock C-Y, Nam A-R, Bang J-H et al (2015) The distinct signatures of VEGF and soluble VEGFR2 increase prognostic implication in gastric cancer. Am J Cancer Res 5:3376–3388

230. Al-Rawi MAA, Watkins G, Mansel RE et al (2005) Interleukin 7 upregulates vascular endothelial growth factor D in breast cancer cells and induces lymphangiogenesis in vivo. Br J Surg 92:305–310

231. Sharma S, Batra RK, Yang SC et al (2003) Interleukin-7 gene-modified dendritic cells reduce pulmonary tumor burden in spontaneous murine bronchoalveolar cell carcinoma. Hum Gene Ther 14:1511–1524

232. Gou HF, Huang J, Shi HS et al (2014) Chemo-immunotherapy with oxaliplatin and interleukin-7 inhibits colon cancer metastasis in mice. PLoS One 9:e85789

233. Zhao YP, Chen G, Feng B et al (2007) Microarray analysis of gene expression profile of multidrug resistance in pancreatic cancer. Chin Med J 120:1743–1752

234. Aoki T, Tashiro K, Miyatake S et al (1992) Expression of murine interleukin 7 in a murine glioma cell line results in reduced tumorigenicity in vivo. Proc Natl Acad Sci U S A 89:3850–3854

235. Sharma S, Wang J, Huang M et al (1996) Interleukin-7 gene transfer in non-small cell lung cancer decreases tumor proliferation, modifies cell surface molecule expression, and enhances antitumor reactivity. Cancer Gene Ther 3:302–313

236. Hock H, Dorsch M, Diamantstein T et al (1991) Interleukin 7 induces CD4 + T cell-dependent tumor rejection. J Exp Med 174:1291–1298

237. McBride WH, Thacker JD, Comora S et al (1992) Genetic modification of a murine fibrosarcoma to produce interleukin 7 stimulates host cell infiltration and tumor immunity. Cancer Res 52:3931–3937

238. Gunnarsson S, Bexell D, Svensson A et al (2010) Intratumoral IL-7 delivery by mesenchymal stromal cells potentiates IFNgamma-transduced tumor cell immunotherapy of experimental glioma. J Neuroimmunol 218:140–144

239. Dwyer CJ, Knochelmann HM, Smith AS et al (2019) Fueling cancer immunotherapy with common gamma chain cytokines. Front Immunol 10:263

240. Sica D, Rayman P, Stanley J et al (1993) Interleukin 7 enhances the proliferation and effector function of tumor-infiltrating lymphocytes from renal-cell carcinoma. Int J Cancer 53:941–947

241. Yuan CH, Yang XQ, Zhu CL et al (2014) Interleukin-7 enhances the in vivo anti-tumor activity of tumor-reactive CD8+ T cells with induction of IFN-gamma in a murine breast cancer model. Asian Pac J Cancer Prev 15:265–271

242. Tang JC, Shen GB, Wang SM et al (2014) IL-7 inhibits tumor growth by promoting T cell-mediated antitumor immunity in Meth A model. Immunol Lett 158:159–166

243. Ditonno P, Too CH, Sakata T et al (1992) Regulatory effects of interleukin-7 on renal tumor infiltrating lymphocytes. Urol Res 20:205–210

244. Lamberti A, Petrella A, Pascale M et al (2004) Activation of NF-kB/Rel transcription factors in human primary peripheral blood mononuclear cells by interleukin 7. Biol Chem 385:415–417

245. Alderson MR, Tough TW, Ziegler SF et al (1991) Interleukin 7 induces cytokine secretion and tumoricidal activity by human peripheral blood monocytes. J Exp Med 173:923–930

246. Ziegler SF, Tough TW, Franklin TL et al (1991) Induction of macrophage inflammatory protein-I f3 gene expression in human monocytes by lipopolysaccharide and IL-7. J Immunol 147:2234–2239

247. Maimela NR, Liu S, Zhang Y (2019) Fates of CD8+ T cells in tumor microenvironment. Comput Struct Biotechnol J 17:1–13

248. Andersson A, Srivastava MK, Harris-White M et al (2011) Role of CXCR3 ligands in IL-7/IL-7Rα-Fc mediated antitumor activity in lung cancer. Clin Cancer Res 17:3660–3672

249. Choi YW, Kang MC, Seo YB et al (2016) Intravaginal administration of Fc-fused IL7 suppresses the cervicovaginal tumor by recruiting HPV DNA vaccine-induced CD8 T cells. Clin Cancer Res 22(23):5898–5908

250. Huang AYC, Golumbek P, Ahmadzadeh M et al (1994) Role of bone marrow-derived cells in presenting MHC class I-restricted tumor antigens. Science (New York, NY) 264:961–965

251. Schroten-Loef C, de Ridder CM, Reneman S et al (2008) A prostate cancer vaccine comprising whole cells secreting IL-7, effective against subcutaneous challenge, requires local GM-CSF for intra-prostatic efficacy. Cancer Immunol Immunother 58:373–381

252. Lee AJ, Zhou X, Chang M et al (2010) Regulation of natural killer T-cell development by deubiquitinase CYLD. EMBO J 29:1600–1612

253. Dubinett SM, Huang M, Dhanani S et al (1993) Down-regulation of macrophage transforming growth factor-beta messenger RNA expression by IL-7. J Immunol 151:6670–6680

254. Sin JI, Kim J, Pachuk C et al (2000) Interleukin 7 can enhance antigen-specific cytotoxic-T-lymphocyte and/or Th2-type immune responses in vivo. Clin Diagn Lab Immunol 7:751–758

255. Tormoen GW, Crittenden MR, Gough MJ (2018) Role of the immunosuppressive microenvironment in immunotherapy. Adv Radiat Oncol 3:520–526

256. Reina M, Espel E (2017) Role of LFA-1 and ICAM-1 in cancer. Cancers (Basel) 9:153

257. Kinter AL, Godbout EJ, McNally JP et al (2008) The common γδ-chain cytokines IL-2, IL-7, IL-15, and

IL-21 induce the expression of programmed death-1 and its ligands. J Immunol 181:6738–6746

258. Bardhan K, Anagnostou T, Boussiotis VA (2016) The PD1:PD-L1/2 pathway from discovery to clinical implementation. Front Immunol 7:550

259. Myklebust JH, Irish JM, Brody J et al (2013) High PD-1 expression and suppressed cytokine signaling distinguish T cells infiltrating follicular lymphoma tumors from peripheral T cells. Blood 121:1367–1376

260. Pfannenstiel LW, Diaz-Montero CM, Tian YF et al (2019) Immune-checkpoint blockade opposes CD8+ T-cell suppression in human and murine cancer. Cancer Immunol Res 7:510–525

261. Veluswamy P, Bruder D (2018) PD-1/PD-L1 pathway inhibition to restore effector functions in exhausted CD8+ T cells: chances, limitations and potential risks. Transl Cancer Res 7(Suppl 4):S530–S537

262. Chen HC, Eling N, Martinez-Jimenez CP et al (2019) IL-7-dependent compositional changes within the γδ T cell pool in lymph nodes during ageing lead to an unbalanced anti-tumour response. EMBO Rep 8:e47379

263. Johnson LDS, Banerjee S, Kruglov O et al (2019) Targeting CD47 in Sézary syndrome with SIRPαFc. Blood Adv 3:1145–1153

264. Eder M, Ottmann OG, Hansen-Hagge TE et al (1990) Effects of recombinant human IL-7 on blast cell proliferation in acute lymphoblastic leukemia. Leukemia 4:533–540

265. Skjønsberg C, Erikstein BK, Smeland EB et al (1991) Interleukin-7 differentiates a subgroup of acute lymphoblastic leukemias. Blood 77:2445–2450

266. Touw I, Pouwels K, van Agthoven T et al (1990) Interleukin-7 is a growth factor of precursor B and T acute lymphoblastic leukemia. Blood 75:2097–2101

267. Dibirdik I, Langlie MC, Ledbetter JA et al (1991) Engagement of interleukin-7 receptor stimulates tyrosine phosphorylation, phosphoinositide turnover, and clonal proliferation of human T-lineage acute lymphoblastic leukemia cells. Blood 78:564–570

268. Canté-Barrett K, Spijkers-Hagelstein JA, Buijs-Gladdines JG et al (2016) MEK and PI3K-AKT inhibitors synergistically block activated IL7 receptor signaling in T-cell acute lymphoblastic leukemia. Leukemia 30:1832–1843

269. Wang H, Zang C, Taing L et al (2014) NOTCH1-RBPJ complexes drive target gene expression through dynamic interactions with superenhancers. Proc Natl Acad Sci U S A 111:705–710

270. González-García S, Toribio ML (2009) Notch 1 signalling in human T-cell development and leukemia. Ther Immunol 28:93–208

271. Tremblay CS, Brown FC, Collett M et al (2016) Loss-of-function mutations of dynamin 2 promote T-ALL by enhancing IL-7 signalling. Leukemia 30:1993–2001

272. Melão A, Spit M, Cardoso BA, Barata JT (2016) Optimal interleukin-7 receptor-mediated signaling, cell cycle progression and viability of T-cell acute lymphoblastic leukemia cells rely on casein kinase 2 activity. Haematologica 101:1368–1379

273. Dalloul A, Laroche L, Bagot M et al (1992) Interleukin-7 is a growth factor for Sézary lymphoma cells. J Clin Invest 90:1054–1060

274. Ge Z, Gu Y, Xiao L et al (2016) Co-existence of IL7R high and SH2B3 low expression distinguishes a novel high-risk acute lymphoblastic leukemia with Ikaros dysfunction. Oncotarget 7:46014–46027

275. Nishii K, Katayama N, Miwa H et al (1999) Survival of human leukaemic B-cell precursors is supported by stromal cells and cytokines: association with the expression of bcl-2 protein. Br J Haematol 105:701–710

276. Scott DW, Gascoyne RD (2014) The tumour microenvironment in B cell lymphomas. Nat Rev Cancer 14:517–534

277. Weitzmann MN, Roggia C, Toraldo G et al (2002) Increased production of IL-7 uncouples bone formation from bone resorption during estrogen deficiency. J Clin Invest 110:1643–1650

278. Giuliani N, Colla S, Morandi F et al (2005) Myeloma cells block RUNX2/CBFA1 activity in human bone marrow osteoblast progenitors and inhibit osteoblast formation and differentiation. Blood 106:2472–2483

279. D'Souza S, del Prete D, Jin S et al (2011) Gfi1 expressed in bone marrow stromal cells is a novel osteoblast suppressor in patients with multiple myeloma bone disease. Blood 118:6871–6880

280. Storti P, Bolzoni M, Donofrio G et al (2013) Hypoxia-inducible factor (HIF)-1α suppression in myeloma cells blocks tumoral growth in vivo inhibiting angiogenesis and bone destruction. Leukemia 27:1697–1706

281. Aldinucci D, Lorenzon D, Olivo K et al (2004) Interactions between tissue fibroblasts in lymph nodes and Hodgkin/Reed-Sternberg cells. Leuk Lymphoma 45:1731–1739

282. Riedel A, Shorthouse D, Haas L et al (2016) Tumor-induced stromal reprogramming drives lymph node transformation. Nat Immunol 17:1118–1127

283. Carrio R, Altman NH, Lopez DM (2009) Downregulation of interleukin-7 and hepatocyte growth factor in the thymic microenvironment is associated with thymus involution in tumor-bearing mice. Cancer Immunol Immunother 58:2059–2072

284. Mandal D, Lahiry L, Bhattacharyya A et al (2006) Tumor-induced thymic involution via inhibition of IL-7R alpha and its JAK-STAT signaling pathway: protection by black tea. Int Immunopharmacol 6:433–444

285. Sorrentino C, Musiani P, Pompa P et al (2011) Androgen deprivation boosts prostatic infiltration of cytotoxic and regulatory T lymphocytes and has no effect on disease-free survival in prostate cancer patients. Clin Cancer Res 17:1571–1581

286. Schroten C, Dits NF, Steyerberg EW et al (2012) The additional value of TGFβ1 and IL-7 to predict

the course of prostate cancer progression. Cancer Immunol Immunother 61:905–910

287. Bednarz-Misa I, Fortuna P, Diakowska D, Jamrozik N, Krzystek-Korpacka M (2020) Distinct local and systemic molecular signatures in the esophageal and gastric cancers: possible therapy targets and bio-markers for gastric cancer. Int J Mol Sci 21:E4509

288. Oka M, Hirose K, Iizuka N et al (1995) Cytokine mRNA expression patterns in human esophageal cancer cell lines. J Interf Cytokine Res 15:1005–1009

289. D'Amico L, Satolli MA, Mecca C et al (2013) Bone metastases in gastric cancer follow a RANKL-independent mechanism. Oncol Rep 29:1453–1458

290. Yamaoka Y, Kita M, Kodama T et al (1995) Expression of cytokine mRNA in gastric mucosa with *Helicobacter pylori* infection. Scand J Gastroenterol 30:1153–1159

291. Yu F, Tian T, Deng B et al (2019) Multi-marker analysis of genomic annotation on gastric cancer GWAS data from Chinese populations. Gastric Cancer 22:60–68

292. Giuntoli RL 2nd, Webb TJ, Zoso A et al (2009) Ovarian cancer-associated ascites demonstrates altered immune environment: implications for anti-tumor immunity. Anticancer Res 29:2875–2884

293. Chang YS, Huang HD, Yeh KT et al (2016) Genetic alterations in endometrial cancer by targeted next-generation sequencing. Exp Mol Pathol 100:8–12

294. Espinoza JA, Jabeen S, Batra R et al (2016) Cytokine profiling of tumor interstitial fluid of the breast and its relationship with lymphocyte infiltration and clinicopathological characteristics. Onco Targets Ther 5:e1248015

295. Vudattu NK, Magalhaes I, Schmidt M et al (2007) Reduced numbers of IL-7 receptor (CD127) expressing immune cells and IL-7-signaling defects in peripheral blood from patients with breast cancer. Int J Cancer 121:1512–1519

296. Midorikawa Y, Tsutsumi S, Taniguchi H et al (2002) Identification of genes associated with dedifferentiation of hepatocellular carcinoma with expression profiling analysis. Jpn J Cancer Res 93:636–643

IL-10 Signaling in the Tumor Microenvironment of Ovarian Cancer

3

Ramesh B. Batchu, Oksana V. Gruzdyn, Bala K. Kolli, Rajesh Dachepalli, Prem S. Umar, Sameer K. Rai, Namrata Singh, Pavan S. Tavva, Donald W. Weaver, and Scott A. Gruber

Abstract

Unlike other malignancies, ovarian cancer (OC) creates a complex tumor microenvironment with distinctive peritoneal ascites consisting of a mixture of several immunosuppressive cells which impair the ability of the patient's immune system to fight the disease. The poor survival rates observed in advanced stage OC patients and the lack of effective conventional therapeutic options have been attributed in large part to the immature dendritic cells (DCs), IL-10 secreting regulatory T cells, tumor-associated macrophages, myeloid-derived suppressor cells, and cancer stem cells that secrete inhibitory cytokines. This review highlights the critical role played by the intraperitoneal presence of IL-10 in the generation of an immunosuppressive tumor microenvironment. Further, the effect of antibody neutralization of IL-10 on the efficacy of DC and chimeric antigen receptor T-cell vaccines will be discussed. Moreover, we will review the influence of IL-10 in the promotion of cancer stemness in concert with the NF-κB signaling pathway with regard to OC progression. Finally, understanding the role of IL-10 and its crosstalk with various cells in the ascitic fluid may contribute to the development of novel immunotherapeutic approaches with the potential to kill drug-resistant OC cells while minimizing toxic side effects.

R. B. Batchu (✉) · O. V. Gruzdyn · S. A. Gruber
Wayne State University School of Medicine, Detroit, MI, USA

John D. Dingell VA Medical Center, Detroit, MI, USA
e-mail: bb9067@wayne.edu; rbatchu@med.wayne.edu; ovgruzdy@med.wayne.edu; scott.gruber@va.gov

B. K. Kolli
Wayne State University School of Medicine, Detroit, MI, USA

John D. Dingell VA Medical Center, Detroit, MI, USA

Med Manor Organics Pvt. Ltd., Hyderabad, India
e-mail: bkolli@wayne.edu; bala.k@medmanor.in

R. Dachepalli · P. S. Umar · S. K. Rai · N. Singh
P. S. Tavva
Med Manor Organics Pvt. Ltd., Hyderabad, India
e-mail: bioscience@medmanor.in; prem@medmanor.in; sameer.rai@medmanor.in; namrata.singh@medmanor.in; tps@medmanor.in

D. W. Weaver
Wayne State University School of Medicine, Detroit, MI, USA
e-mail: dweaver@med.wayne.edu

A. Birbrair (ed.), *Tumor Microenvironment*, Advances in Experimental Medicine and Biology 1290, https://doi.org/10.1007/978-3-030-55617-4_3

Keywords

Ovarian cancer · Tumor microenvironment ·
Immunosurveillance · Cytokines ·
Interleukins · Interleukin-10 · Transforming
growth factor -β · Regulatory T cells ·
Dendritic cells · Immunotherapy · Nuclear
factor-κB · Cancer stem cells · Epithelial-to-
mesenchymal transition · CD4+ T cells ·
CD8+ cytotoxic T cells · Janus kinase-signal
transducer and activator of transcription 3
pathway · Interferon-γ · Interleukin-12 ·
Chemokines · Chimeric antigen receptor T
cells · Mesothelin

3.1   Introduction

Ovarian cancer (OC) remains the leading cause
of mortality from gynecologic malignancies,
with an estimated 22,530 new cases and 13,980
deaths in the US in 2020 [1]. The majority of
cases are advanced International Federation of
Gynecology and Obstetrics stage III or IV at the
time of diagnosis, thus precluding surgical debulking [2]. Accumulation of peritoneal ascites, a
heterogeneous fluid that harbors a wide variety of
cellular and soluble factors which promote survival pathways in tumor cells while inducing
immunosuppression, is an important manifesta-
tion of advanced OC [3]. Although paclitaxel and
platinum-based combinatorial chemotherapy in
addition to surgical tumor debulking achieve an
initial clinical response in the majority of patients,
tumor relapse due to peritoneal accumulation of
inflammatory ascites remains a major problem
[4]. Peritoneal ascitic fluid is a complex mixture
of cellular components and soluble growth factors, including but not limited to cytokines such
as interleukin (IL)-10 and transforming growth
factor (TGF)-β, creating an inflammatory microenvironment that is conducive to the development of chemo-resistance and tumor growth [3].
Along these lines, lack of effective treatment for
OC can be in large part attributed to a poor understanding of the peritoneal ascites-associated
tumor microenvironment.

The healthy immune system has professional
antigen (Ag)-presenting cells, such as dendritic
cells (DCs), that endocytose dying tumor cells,
process them into peptides, and express the peptides on their cell surface in conjunction with
major histocompatibility (MHC) class I Ag. DCs
then migrate to secondary lymphoid organs
where they activate naïve CD8+ T cells against
tumor peptides to generate cytotoxic T cells
which lyse tumor cells [5, 6], a process termed
immunosurveillance [7]. This ability of the
healthy immune system to detect inflammatory
pre-cancerous lesions and eliminate tumor cells
is often known as the *elimination phase* [8].

Immunosurveillance is predominant in the initial stages of OC, before the composition of the
immune infiltrate changes from an anti- to
pro-tumorigenic immunotolerant leukocyte profile with the onset of tumor progression [9].
Immune checkpoints, such as programmed cell
death 1 and cytotoxic T-lymphocyte-associated
protein 4, are cellular hard-wired pathways that
ward off autoimmunity during immune responses
against various microbial agents [10]. However,
the growing tumor's ability to cause chronic
inflammation and camouflage itself from immunosurveillance co-opts these immune checkpoints, further dampening the activity of both
DCs and T cells [11]. At this stage, a significant
upregulation of IL-10 expression is observed in
OC patients [12]. This process results in the
enhanced interplay of the patient's immune system with neoplastic cells in the peritoneal cavity
leading to the development of an inhibitory
microenvironment conducive to further growth
and proliferation of OC cells, often called the
equilibration phase [13].

The core of the proliferating tumor consists of
cancer stem cells (CSCs), a subpopulation that
has the capability for self-renewal and pluripo-
tency that promotes tumor recurrence and metastasis [14]. At this stage, the tumor tissue starts
developing a distinct but interdependent second
compartment called the stroma that includes con-
nective tissue, blood vessels, extracellular matrix
proteins, and various infiltrating immune cells.
The infiltrates include immature DCs, tumor-
associated macrophages, regulatory T cells,

myeloid-derived suppressor cells, and cancer-associated fibroblasts, among others [15]. Stem cells in OC are known to induce the M2 polarization of macrophages influenced by IL-10 secretion [16]. CSCs and the corresponding stroma co evolve over time, producing an environment conducive to more potent tumor growth while imposing a barrier to immunosurveillance [17]. Indeed, CSCs that escape chemotherapy are the main culprits for the spread of disease [18]. Various immunosuppressive bioactive compounds such as polyamines, vascular endothelial growth factor, IL-10, and TGF-β are secreted into this tumor microenvironment [14]. At this stage, non-motile, epithelial OC cells gain mesenchymal-like migratory and invasive properties, a process known as epithelial-to-mesenchymal transition (EMT) which is a hallmark in OC progression. Cells break free from the primary tumor and spread to other organs via blood and lymphatic vessels, a metastatic process often referred to as the *escape phase* [8, 19, 20].

Immunotherapies have had limited success in controlling OC due to its ability to create an immunosuppressive milieu within the peritoneal cavity which inhibits not only the intratumoral migration of cytotoxic T cells, but also allows substantial infiltration of inhibitory cellular and soluble factors [21]. Prominent among these are regulatory T cells and their secreted cytokines [22, 23]. The role of IL-10 in the formation of immunosuppressive peritoneal ascites and its contribution to ovarian carcinogenesis is still an active area of investigation [24–26]. Certainly, a better understanding of the role of IL-10 in the pathogenesis of OC would contribute to the development of effective cell-based therapies.

3.2   DCs, T Cells, and Cell-Mediated Immune Responses

DC- and T-cell-based cellular immunotherapies have improved patient survival; however, tumor relapse is common due to various immunosuppressive cytokines present in peritoneal ascites

[27]. Two types of DC precursors, conventional/myeloid, and plasmacytoid DCs have been identified. Upon activation, the former secretes high levels of IL-12 and stimulates naïve CD4+ T cells to differentiate into Th1 cells which facilitate strong, cell-mediated immunity converting CD8+ naïve T cells into cytotoxic T cells. In contrast, the latter often present at the site of the tumor and produce low levels of IL-12 and induce the development of Th2 cells which suppress cell-mediated immunity. IL-10 was also shown to promote the anti-apoptotic activity of malignant ascites [28, 29], and its expression consistently correlates with a predominantly Th2 response in advanced stage OC [30, 31]. A host of other CD4+ Th subsets such as Th17 have been identified that modulate CD8+ effector T-cell activity, but their exact role in the immune response is still being elucidated [32].

3.3   Interleukins in the OC Microenvironment

Interleukins are small, signaling proteins which belong to a broad category of cytokines that also include chemokines, interferons (IFNs), and lymphokines [33]. Interleukins are synthesized by a majority of immune cells that are significant mediators of immunoregulatory pathways, initiate signal transduction with their cell-surface receptor engagement, and are classified into pro- or anti-inflammatory subsets [33]. Multiple interleukins are associated with OC pathogenesis, and the balance between pro- and anti-inflammatory cytokine profiles within the peritoneal stroma will determine the degree of proliferation and invasiveness of OC cells [34, 35].

3.4   IL-10 in the OC Microenvironment

IL-10 is a potent, anti-inflammatory cytokine initially referred to as cytokine synthesis inhibitory factor [36] and plays a significant role in the pathogenesis of OC [37]. It is secreted by a majority of innate immune cells, including monocytes, DCs, macrophages, and natural killer cells; adaptive immune cells, such as CD4+ T cells, CD8+ T cells, Th17 cells, and B cells; and OC cells themselves [23, 38, 39]. It is well established that the role of IL-10 in the tumor microenvironment is to facilitate shielding from immunosurveillance [40, 41]. High levels of IL-10 are found in both the serum and peritoneal effusions of patients with serous OC, particularly those with advanced stage disease [23, 31, 42]. IL-10 is known to differentiate naïve CD4+ T cells into Th2 cells which suppress adaptive immunity [30]. Enhanced IL-10 secretion is also observed at the time of EMT, promoting the loss of E-cadherin and acquisition of N-cadherin which increases tumor cell motility [43, 44]. In fact, IL-10-neutralizing antibodies have been shown to markedly reverse this transition [45]. Further, a positive correlation of IL-10 expression with cell migration from the peritoneal cavity has been reported, thus corroborating its role in EMT and promotion of metastasis [37].

Akin to its ligand, IL-10 receptor expression has been observed in macrophages, T cells, and DCs [40]. OC cells expressed higher levels of IL-10 receptor than the surrounding stroma, suggesting that these receptors may be involved in the pathogenesis of the disease [23]. IL-10 gene expression is finely orchestrated by regulatory transcriptional networks (Fig. 3.1). Engagement of IL-10 with its receptor on the cell surface results in the recruitment and activation of Janus kinase 1 (JAK1) and tyrosine kinase 2 (TYK2) at the cytoplasmic end [46]. This complex then recruits and phosphorylates signal transducer and activator of transcription 3 (STAT3), facilitating its translocation into the nucleus [46]. This translocation leads to target gene expression of various pro-tumorigenic and apoptosis resistance genes [47]. An important target gene of STAT3 is Twist, which is key in facilitating EMT and promoting invasiveness and metastasis [48].

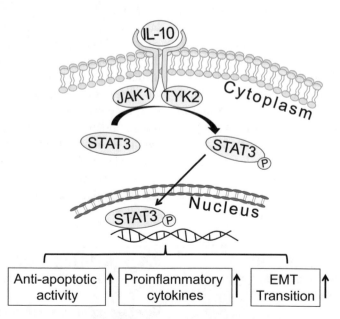

Fig. 3.1 Schematic representation of the IL-10 signaling pathway: ligation of IL-10 with its receptor results in the JAK1/TYK2-mediated phosphorylation of STAT3 followed by its nuclear translocation. This translocation leads to the expression of anti-apoptotic and pro-tumorigenic genes contributing to EMT. *JAK1* Janus kinase 1, *TYK2* Non-receptor tyrosine-protein kinase, *STAT3* Signal transducer and activator of transcription 3, *EMT* epithelial-mesenchymal transition

3.5   Cell-Based Immunotherapy for OC to Overcome the Inhibitory Tumor Microenvironment

Cell-based immunotherapy is a promising approach to OC [49] based on the strong positive correlation observed between tumor infiltrating lymphocytes and improved overall survival [50]. The rejection of metastatic tumors in patients following infusion of autologous, tumor-specific CD8+ cytotoxic T cells further strengthens this hypothesis [51]. Although the CD8+ adaptive arm of the immune response uses cytotoxic T cells as the main weapon to lyse tumor cells, CD4+ naïve T cells are required for optimal cytotoxic T-cell activity [52]. Despite successful demonstration of in vitro tumor cell killing by ex vivo generated DCs, clinically meaningful responses have been sparse in DC vaccine trials due to the inhibition of DC maturation and function by the recruitment of immunosuppressive cells and cytokines into the tumor site in vivo [53]. One way to circumvent this is to generate DCs ex vivo, expose them to tumor-specific Ag, and subsequently administer this preparation to patients as a therapeutic cancer vaccine. DCs would then be expected to migrate to secondary lymphoid organs and induce robust cytotoxic T-cell responses. Cellular immunotherapies with tumor Ag-pulsed DCs and ex vivo activation of T cells for adoptive transfer improved patient survival [54, 55].

3.6   Influence of IL-10 on DC-Mediated Immune Responses

Tumor cells themselves are potential immunogens but are generally not capable of initiating therapeutically useful immune responses [51]. DCs are central to shaping the adaptive immune response as professional Ag-presenting cells, and modulate the response against tumor Ags leading to clonal expansion of CD8+ naïve T cells into cytotoxic T cells [6]. DCs are found in most tumors and acquire tumor Ags either by 'nibbling' live or cap-turing dying tumor cells. Tumor Ags are processed into peptides that are expressed on the cell surface of DCs in conjunction with MHC class I Ags. DCs then migrate to lymph nodes, where they prime CD8+ naïve T cells for clonal expansion into cytotoxic T cells. This process required interaction of the MHC class I-tumor peptide complex on DCs with the CD8+ T-cell receptor as well as the interaction of CD80/CD86 co-stimulatory molecules on DCs with CD8+ T cells. IL-10 production is also known to influence the percentage of DCs in OC patients [56].

Often, substantial numbers of DCs are observed in surrounding tissue, but their infiltration into the tumor bed is limited. Even when they make their way into the tumor, DCs and T cells do not function effectively due to tumor-induced immunosuppression [51]. Studies have revealed that the maturation and function of DCs are adversely affected by the IL-10-rich tumor microenvironment, causing the activation of the JAK-STAT3 pathway [57–59]. This pathway results in the loss of DC tumor Ag expression, thus limiting its presentation to T cells contributing to inhibition of cell-mediated immunity. This is one of the ways IL-10 is able to facilitate evasion of OC cells from immune recognition, and studies have reported a similar IL-10-induced downregulation of MHC class I in OC cells [60, 61]. Increased serum levels of IL-10 significantly correlated with not only numerical deficiencies of DCs, but also with increased circulation of immature phenotypes of circulating DC subsets as well as impairment of DC differentiation. IL-10 inhibited anti-tumor immunity by inducing regulatory T cells and inhibited DC tumor Ag presentation to impair activation of CD4+ helper and CD8+ cytotoxic T cells [62, 63]. In this regard, IL-10 has been implicated in blocking proliferation and cytokine production and is known to play a role in inducing T-cell anergy [64], thus shielding tumor cells from immunosurveillance.

In order to improve outcomes with DC-based therapeutic cancer vaccines, it is necessary to block the immunosuppressive environment while at the same time provide the means for effective cell-mediated cytotoxic immune responses against tumor cells. Of several OC immunotherapeutic approaches currently under development, we will focus on the ex vivo generation of DC vaccines [65–71] and chimeric antigen receptor (CAR) T-cell therapies [72–76] that are currently being studied in our laboratory.

3.7 Efficient Tumor Ag Delivery and Inhibitory Tumor Microenvironment: Key Factors that Influence DC Immunotherapy

Immunotherapy is a particularly appealing approach due to its potential for eliminating tumor cells that are often unreachable by conventional therapies with negligible side effects. Despite successful elimination of tumor cells in vitro, clinically meaningful responses in DC vaccine trials have been sparse, due to both the suboptimal intracellular bioavailability of tumor Ag as well as the inhibitory tumor microenvironment. Melanoma associated antigen A3 (MAGE-A3) is a tumor-specific Ag that belongs to the family of cancer testis Ags that are restricted to tumor cells and immune-privileged gonadal germ cells. It has attracted particular attention as a potential target for immunotherapy since it is more highly expressed in advanced cancer stages and its presence is associated with a poorer prognosis. MAGE-A3 expression has been observed in various OC cell lines [77] and is positively correlated with disease status in a large number of OC specimens [78]. Efficient intracellular tumor Ag delivery to access MHC class I molecules in the cytoplasm is essential in order to generate robust cell-mediated responses and produce tumor cell lysis.

3.8 Efficient Tumor Ag Delivery with Cell-Penetrating Domains (CPDs)

First-generation immunotherapies loading DCs with tumor-associated antigens, tumor lysates, tumor RNA, or tumor peptides all showed limited success due to the impervious nature of the DC cell membrane to large tumor protein Ag, resulting in poor presentation of peptides to human leukocyte antigen (HLA) class I molecules [55]. However, various synthetic cell penetrating domains (CPDs) are known to ferry covalently linked heterologous Ags to the intracellular compartment by traversing the plasma membrane [70, 79, 80], a process known as protein transduction [81]. CPDs are non-immunogenic, short amino-acid motifs based on the sequence identified in HIV-1 transactivator of transcription protein. Tumor Ag expressed in-frame with CPD can penetrate through the plasma membrane to directly enter the cytosol, thus gaining access to the HLA class I pathway [81, 82].

We cloned and purified MAGE-A3 in-frame with CPD and documented enhanced cytosolic bioavailability in DCs without altering cell functionality [71] when compared with existing MAGE-A3 protein alone [83] and peptide vaccines [84]. We observed very little fluorescent staining in MAGE-A3-pulsed DCs even after 2 h, but in contrast, CPD-MAGE-A3 penetrated the DCs within 5 min after pulsing [65]. Studies using deconvolution and confocal microscopy confirmed that CPD-MAGE-A3 was localized to the DC cytosol, clearly demonstrating a rapid and efficient way to introduce tumor Ag into the cytoplasm [70, 71] with more effective anti-tumor cell-mediated responses [65].

3.9 Efficient Tumor Ag Delivery with Recombinant Adeno-Associated Viral (rAAV) Vectors

Bioengineered recombinant adeno-associated viral (rAAV) vectors represent another approach to effectively present tumor Ag intracellularly to DCs, and are promising for human gene therapy because of their excellent safety profile, including a modest frequency of integration, minimal generation of immune responses, and lack of association with human disease [85]. Along these lines, rAAV-2 vectors have been shown to transduce DCs efficiently followed by tumor cell lysis

[86]. Further, rAAV-2 pseudo-typed virus with an adeno-associated virus type 6 (AAV-6) capsid exhibited a high degree of tropism for DCs [87], resulting in enhanced generation of cytotoxic T-lymphocytes [88]. We demonstrated enhanced the cytosolic bioavailability of MAGE-A3 with targeted killing of OC with MAGE-A3 cytotoxic T cells by two different approaches. Our results indicate that DCs can elicit effective antitumor responses against MAGE-A3-expressing OC cell lines, and provide a basis for potential translation to the clinical arena [69].

3.10 Use of Human OC Cell Line SKOV-3-Conditioned Medium to Simulate Peritoneal Malignant Ascites

Although inadequate cytoplasmic expression of tumor Ag can be addressed by using CPDs, rAAV anti-tumor immunity is substantially impaired by inhibitor cytokines such as TGF-β and IL-10 in the tumor microenvironment. Since OC spreads primarily via the peritoneal cavity, malignant ascites may be an ideal fluid in which to unravel the effects of the tumor microenvironment on cell-mediated immunity. Malignant ascites impairs DC functionality in a variety of ways, such as inducing morphological changes of short and reduced dendrites, decreasing expression of co-stimulatory molecules, and inducing Ag-specific anergy of cytotoxic T cells [89–94]. Understanding the immunosuppression mediated by malignant ascites may facilitate the development of effective DC immunotherapy. Along these lines, we have used conditioned medium from the human OC cell line SKOV-3 to simulate malignant ascites [66, 68, 72, 75]. The SKOV-3 culture medium was replaced with serum-free medium at approximately 60% confluency and the tumor-conditioned medium was harvested after 48 h. DCs were generated from the adherent fraction of peripheral blood mononuclear cells to study the effects of the tumor microenvironment.

We observed that typical mature DCs morphologically displayed large sizes and irregular shapes with dendrites protruding from the cells [65, 67, 71]. These DCs had high expression of HLA-DR, CD40, CD80 (B7.1) and CD86 (B7.2), which are T-cell co-stimulatory molecules required for the activation of specific effector CD4+ and CD8+ T cells against cancer cells. Along these lines, shorter and fewer dendrites were observed in our morphologic studies of ex vivo generated DCs treated with conditioned medium thereby restricting dendritic interaction with naïve T cells (Fig. 3.2). Advanced OC is manifested by the decreased expression of DC co-stimulatory molecules (e.g., CD80, CD83, and CD86) with Ag-specific anergy of cytotoxic T cells [95–97]. Corroborating these reports, we observed downmodulation of DC co-stimulatory molecules CD80 and CD86 when DCs were exposed to OC-conditioned medium (Fig. 3.3).

3.11 Influence of Malignant Ascites on DC Migration

Chemokine networks play a crucial role in the homing of DCs and T cells as well as in regulating the spread of cancer cells to distant organs. Chemokine receptor (CCR) 7 normally facilitates the migration of mature DCs toward chemokine ligands (CCL) such as CCL19, CCL5, and CCL21 on naïve T cells in secondary lymphoid organs, an important event for cell-mediated immune response with cytotoxic T-cell generation [98]. In fact, advanced OC is also characterized by the downmodulation of chemokine molecules (e.g., CCR7, CCL21) necessary to direct the traffic of DCs to lymph nodes [99, 100]. We documented that DCs exposed to OC-conditioned medium showed reduced expression of CCR7 by flow cytometric and reverse transcriptase polymerase chain reaction (RT-PCR) analysis (Fig. 3.4a, b). Addition of OC-conditioned medium resulted in a 40% inhibition of DC migration toward CCL21 as well as downregulation of CCR7 expression levels (Fig. 3.5a) [67].

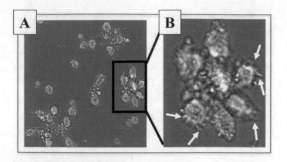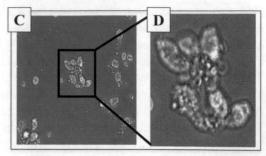

Fig. 3.2 Altered morphology and dendrites of DC in OC-conditioned medium: dendrites characteristic of mature DC are apparent in controls (**A, B**) but shorter and fewer in OC cell line (SKOV-3)-conditioned medium (**C, D**). **A** and **C**, 40x; **B** and **D**: 200x magnification. Arrows pointing dendrites

Fig. 3.3 Flow-cytometric detection of DC co-stimulatory molecules CD80 and CD86: reduced levels of both CD80 and CD86 were observed in the presence of OC cell line (SKOV-3)-conditioned medium

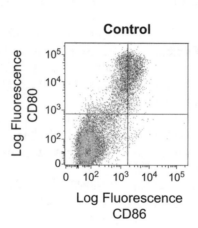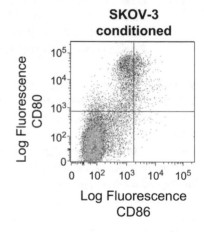

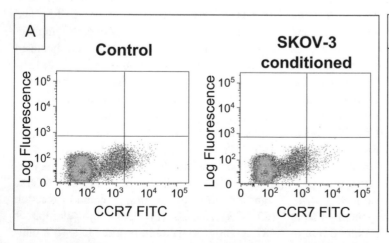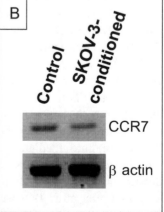

Fig. 3.4 Expression of CCR7 in OC-conditioned medium: (**A**) Flow cytometric analysis of CCR7 on the cell surface using CCR7 antibody labeled with FITC. (**B**) RT-PCR analysis of CCR7 mRNA expression in DCs. mRNA from DCs was isolated followed by first-strand cDNA synthesis and agarose gel analysis

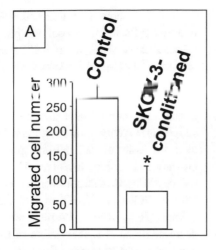
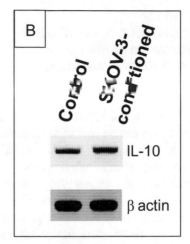

Fig. 3.5 Migration of DCs and expression of IL-10 in OC-conditioned medium: (**A**) DC migration toward CCL19 in Boyden chamber experiments compared with control (*$p < 0.05$). Error bars represent mean ± SD for three separate experiments. (**B**) RT-PCR analysis of CCR7 mRNA expression in DCs. mRNA from DCs was isolated followed by first-strand cDNA synthesis and agarose gel analysis

3.12   IL-10 Blockade Reverses Immune Inhibitory Effects of the Tumor Microenvironment on DCs

IL-10 potently inhibits the production of pro-inflammatory cytokines such as IFN-γ, tumor necrosis factor (TNF)-α, and IL-12 in DCs with a significant reduction in the expression of MHC class II molecules [101]. Further, IL-10 blocks the upregulation of these molecules, thus impairing the ability of DCs to stimulate T-cell responses and further inducing a state of Ag-specific tolerance. Earlier studies indicated IL-10-mediated inhibition of DC maturation and function via reduced expression of MHC and co-stimulatory molecules (e.g., CD40, CD80, and CD86) [102, 103]. Indeed, we observed upregulation of IL-10 when DCs were cultured in OC-conditioned medium (Fig. 3.5b).

IL-10 decreases the production of pro-inflammatory cytokines and has been shown to inhibit the expression of CCR7 [101]. Along these lines, IL-10-mediated increase of the pro-inflammatory potential of CCL5 in the tumor microenvironment has been observed via sensory nerves [104]. In fact, cancer progression is inhibited by the innervation of sensory nerves within the tumor microenvironment [105]. These results corroborate earlier observations that intratumoral administration of CCL21 reduced the growth of distant metastases by enhancing immune cell infiltration into the tumor bed [106].

Studies from our laboratory and others suggest that inhibition of IL-10 may enhance the immune response and clinical benefit derived from DC vaccination in OC patients [56, 68, 102, 107, 108]. We observed that blocking IL-10 and TGF-β with specific-antibody restored expression levels of DC co-stimulatory molecules and migration toward the CCL21 chemokine in trans-well experiments with tumor-conditioned medium [67, 68]. We made similar observations in DCs generated with rAAV delivery of MAGE-A3 tumor-specific Ag, where anti-tumor T-cell responses in tumor-conditioned medium were enhanced by blocking IL-10 and TGF-β in vitro [68, 69]. Significant reversal of tumor-derived immunosuppression may be achieved by blocking IL-10 and TGF-β in the microenvironment of OC and pancreatic cancer [68], potentially allowing for development of more effective vaccines.

3.13 OC Tumor Microenvironment-Induced Modulation of the NF-κB Signaling Pathway in DCs and T-Cell Interactions

Nuclear factor (NF)-κB, a heterodimer of p50 and p65 subunits, is a natural part of the immune defense that provides survival signals for DCs and T cells. However, OC cells subvert the NF-κB signaling pathway to orchestrate chemoresistance, metastasis, immune evasion, and tumor progression. IL-10 is known to promote cancer stemness via the NF-κB pathway contributing to metastasis [109]. The activation of canonical NF-κB signaling has been well established in anti-apoptotic and immunomodulatory functions in response to the tumor microenvironment, while the non-canonical pathway is important in cancer stem cell maintenance and tumor re-initiation [110, 111]. NF-κB activity in OC helps to create an immune-evasive environment and enhance expression of tumor-promoting cytokines [110]. We examined the NF-κB pathway in the OC-induced microenvironment in order to delineate strategies for restoring immunocompetence, thereby enhancing cancer cell killing.

NF-κB is bound to and sequestered by the inhibitor IκBα in the cytosol, but phosphorylation by IκBα kinase (IKK) at the Ser32 residue of IκBα renders it inactive and releases NF-κB. The IKK complex, in turn, must first be activated by phosphorylation at its Ser176 residue. NF-κB then translocates into the nucleus with enhanced transcriptional activity. Our laboratory further examined the role of the NF-κB pathway in the OC-induced microenvironment. Addition of OC-conditioned medium decreased expression of both of IKKα and IKKβ in DC and T-cell cocultures (Fig. 3.6A). Further, we observed a decrease in the phosphorylation of the IKK complex at Ser176, indicating its inactivation. Moreover, expression of IκBα was increased along with a concomitant reduction in Ser32 phosphorylation, correlating with both IKK inactivation and decreased expression of the NF-κB p65 subunit (Fig. 3.6B). Thus, using OC-conditioned medium, we demonstrated increased presence of active IκBα which binds and sequesters NF-κB, preventing its translocation into the nucleus ultimately inhibiting DC activation [66]. Knowledge of this NF-κB pathway may prove useful in the subsequent design of DC immunotherapeutic protocols to counter the immunosuppressive nature of the tumor microenvironment.

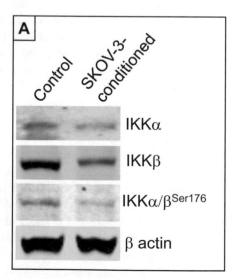

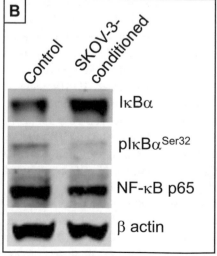

Fig. 3.6 Expression and phosphorylation of whole cell lysates of mock- and SKOV-3 conditioned DCs were immunoblotted with monoclonal antibody for indicated proteins upstream of (**A**) and downstream from (**B**) the NF-kB signaling pathway. β actin was used as the loading control

3.14 CAR T-Cell Therapy to Overcome Immunosuppression

CAR T cell therapy is considered a radical departure from existing treatments since it bypasses DC processing and presentation of tumor Ag in conjunction with MHC, which is often dysfunctional in tumors, and combines antibody specificity to recognize tumors. Thus, CAR T-cell therapy improves upon earlier cell-based cancer immunotherapies in that it overcomes the negative effects of the immunosuppressive tumor microenvironment. CARs are comprised of an extracellular single-chain fragment variable derived from antibody against tumor-specific Ag and connected to intracellular co-stimulatory and activation domains.

We isolated human T cells from blood drawn from healthy individuals and transduced them with third-generation mesothelin-CAR lentiviral vector to generate mesothelin CAR T cells. Mesothelin CAR T cells, upon contact with SKOV-3 cells in vitro, induced secretion of granzyme B and IFN-γ, both of which are crucial for rapid initiation of cytotoxicity with an efficient lysis of OC cells [76]. Addition of tumor-conditioned medium simulating the tumor microenvironment significantly inhibited IFN-γ and granzyme B secretion. Their levels were partially restored to baseline when tumor-conditioned medium was IL-10 depleted. We also observed a reduction in the cytotoxicity of mesothelin CAR T cells against SKOV-3 cells in the presence of conditioned medium and blunting of this effect with IL-10 depletion. Our observations indicate that a significant reversal of tumor-derived immunosuppression may be achieved by blocking IL-10 in the local microenvironment, allowing for more effective cytotoxicity of mesothelin-engrafted CAR T cells [74].

Although lentiviral-mediated generation of CAR T cells is convenient, these vectors pose safety concerns. We incorporated Sleeping Beauty and mini-circle design enhancements into IL-2-secreting natural NK-92MI cells to eliminate both bacterial and viral components and address inhibition by the tumor microenviron-

ment. The hybrid Sleeping Beauty and minicircle technologies provided increased engraftment and cytotoxicity in vitro [73]. We observed a significant reversal of tumor microenvironment-mediated inhibition of mesothelin-CAR T-cell activity either by depleting IL-10 in autologous T cells or by providing IL-2. The ability of IL-2 addition to kill OC cells was further enhanced by depleting IL-10 from the tumor microenvironment [72].

In summary, the limited success of prior immunotherapeutic approaches has been attributed primarily to co-opting of immune checkpoints with increased secretion of immunosuppressive cytokines via modulation of NF-κB inflammatory signaling pathways [112]. We posit that blocking either regulator T cells or their secretion of immune inhibitory cytokines such as IL-10 in the OC microenvironment will allow for more effective translation of potential immunotherapies into the clinical setting.

References

1. Siegel RL, Miller KD, Jemal A (2020) Cancer statistics, 2020. CA Cancer J Clin 70(1):7–30
2. Babayeva A, Braicu EI, Grabowski JP et al (2018) Clinical outcome after completion surgery in patients with ovarian cancer: the Charite experience. Int J Gynecol Cancer 28(8):1491–1497
3. Carduner L, Leroy-Dudal J, Picot CR, Gallet O, Carreiras F, Kellouche S (2014) Ascites-induced shift along epithelial-mesenchymal spectrum in ovarian cancer cells: enhancement of their invasive behavior partly dependent on alphav integrins. Clin Exp Metastasis 31(6):675–688
4. Yan L, Anderson GM, DeWitte M, Nakada MT (2006) Therapeutic potential of cytokine and chemokine antagonists in cancer therapy. Eur J Cancer 42(6):793–802
5. Trumpfheller C, Longhi MP, Caskey M et al (2012) Dendritic cell-targeted protein vaccines: a novel approach to induce T-cell immunity. J Intern Med 271(2):183–192
6. Steinman RM (2012) Decisions about dendritic cells: past, present, and future. Annu Rev Immunol 30:1–22
7. Chan CW, Housseau F (2008) The 'kiss of death' by dendritic cells to cancer cells. Cell Death Differ 15(1):58–69
8. Dunn GP, Bruce AT, Ikeda H, Old LJ, Schreiber RD (2002) Cancer immunoediting: from immunosurveillance to tumor escape. Nat Immunol 3(11):991–998

9. Hamanishi J, Mandai M, Konishi I (2016) Immune checkpoint inhibition in ovarian cancer. Int Immunol 28(7):339–348

10. Doo DW, Norian LA, Arend RC (2019) Checkpoint inhibitors in ovarian cancer: a review of preclinical data. Gynecol Oncol Rep 29:48–54

11. Pardoll DM (2012) The blockade of immune checkpoints in cancer immunotherapy. Nat Rev Cancer 12(4):252–264

12. Li L, Ma Y, Xu Y (2019) Follicular regulatory T cells infiltrated the ovarian carcinoma and resulted in CD8 T cell dysfunction dependent on IL-10 pathway. Int Immunopharmacol 68:81–87

13. Norling L, Serhan C (2010) Profiling in resolving inflammatory exudates identifies novel anti-inflammatory and pro-resolving mediators and signals for termination. J Intern Med 268(1):15–24

14. Magee JA, Piskounova E, Morrison SJ (2012) Cancer stem cells: impact, heterogeneity, and uncertainty. Cancer Cell 21(3):283–296

15. Gajewski TF, Schreiber H, Fu YX (2013) Innate and adaptive immune cells in the tumor microenvironment. Nat Immunol 14(10):1014–1022

16. Deng X, Zhang P, Liang T, Deng S, Chen X, Zhu L (2015) Ovarian cancer stem cells induce the M2 polarization of macrophages through the PPARgamma and NF-kappaB pathways. Int J Mol Med 36(2):449–454

17. Vesely MD, Kershaw MH, Schreiber RD, Smyth MJ (2011) Natural innate and adaptive immunity to cancer. Annu Rev Immunol 29:235–271

18. House CD, Hernandez L, Annunziata CM (2015) In vitro enrichment of ovarian cancer tumor-initiating cells. J Vis Exp 96

19. Syn N, Wang L, Sethi G, Thiery JP, Goh BC (2016) Exosome-mediated metastasis: from epithelial-mesenchymal transition to escape from immunosurveillance. Trends Pharmacol Sci 37(7):606–617

20. Arum CJ, Anderssen E, Viset T et al (2010) Cancer immunoediting from immunosurveillance to tumor escape in microvillus-formed niche: a study of syngeneic orthotopic rat bladder cancer model in comparison with human bladder cancer. Neoplasia 12(6):434–442

21. Wang W, Kryczek I, Dostal L et al (2016) Effector T cells abrogate stroma-mediated chemoresistance in ovarian cancer. Cell 165(5):1092–1105

22. Jammal MP, Martins-Filho A, Silveira TP, Murta EF, Nomelini RS (2016) Cytokines and prognostic factors in epithelial ovarian cancer. Clin Med Insights Oncol 10:71–76

23. Rabinovich A, Medina L, Piura B, Huleihel M (2010) Expression of IL-10 in human normal and cancerous ovarian tissues and cells. Eur Cytokine Netw 21(2):122–128

24. Krishnan V, Berek JS, Dorigo O (2017) Immunotherapy in ovarian cancer. Curr Probl Cancer 41(1):48–63

25. Wang DH, Guo L, Wu XH (2015) Checkpoint inhibitors in immunotherapy of ovarian cancer. Tumour Biol 36(1):33–39

26. Schwab CL, English DP, Roque DM, Pasternak M, Santin AD (2014) Past, present and future targets for immunotherapy in ovarian cancer. Immunotherapy 6(12):1279–1293

27. Ghoneum A, Afify H, Salih Z, Kelly M, Said N (2018) Role of tumor microenvironment in ovarian cancer pathobiology. Oncotarget 9(32):22832–22849

28. Roy T, Paul S, Baral RN, Chattopadhyay U, Biswas R (2007) Tumor associated release of interleukin-10 alters the prolactin receptor and down-regulates prolactin responsiveness of immature cortical thymocytes. J Neuroimmunol 186(1–2):112–120

29. Hagemann T, Robinson SC, Thompson RG, Charles K, Kulbe H, Balkwill FR (2007) Ovarian cancer cell-derived migration inhibitory factor enhances tumor growth, progression, and angiogenesis. Mol Cancer Ther 6(7):1993–2002

30. Dobrzanski MJ, Rewers-Felkins KA, Samad KA et al (2012) Immunotherapy with IL-10- and IFN-gamma-producing CD4 effector cells modulate "natural" and "inducible" CD4 TReg cell subpopulation levels: observations in four cases of patients with ovarian cancer. Cancer Immunol Immunother 61(6):839–854

31. Mustea A, Braicu EI, Koensgen D et al (2009) Monitoring of IL-10 in the serum of patients with advanced ovarian cancer: results from a prospective pilot-study. Cytokine 45(1):8–11

32. Zhu X, Ma D, Zhang J et al (2010) Elevated interleukin-21 correlated to Th17 and Th1 cells in patients with immune thrombocytopenia. J Clin Immunol 30(2):253–259

33. Smyth MJ, Cretney E, Kershaw MH, Hayakawa Y (2004) Cytokines in cancer immunity and immunotherapy. Immunol Rev 202(1):275–293

34. Yigit R, Figdor CG, Zusterzeel PL, Pots JM, Torensma R, Massuger LF (2011) Cytokine analysis as a tool to understand tumour-host interaction in ovarian cancer. Eur J Cancer 47(12):1883–1889

35. Bushley AW, Ferrell R, McDuffie K et al (2004) Polymorphisms of interleukin (IL)-1alpha, IL-1beta, IL-6, IL-10, and IL-18 and the risk of ovarian cancer. Gynecol Oncol 95(3):672–679

36. Moore KW, Vieira P, Fiorentino DF, Trounstine ML, Khan TA, Mosmann TR (1990) Homology of cytokine synthesis inhibitory factor (IL-10) to the Epstein-Barr virus gene BCRFI. Science 248(4960):1230–1234

37. Lane D, Matte I, Garde-Granger P, Bessette P, Piche A (2018) Ascites IL-10 promotes ovarian cancer cell migration. Cancer Microenviron 11(2–3):115–124

38. Mannino MH, Zhu Z, Xiao H, Bai Q, Wakefield MR, Fang Y (2015) The paradoxical role of IL-10 in immunity and cancer. Cancer Lett 367(2):103–107

39. Zhou J, Ye F, Chen H, Lv W, Gan N (2007) The expression of interleukin-10 in patients with primary ovarian epithelial carcinoma and in ovarian carcinoma cell lines. J Int Med Res 35(3):290–300

40. Sato T, Terai M, Tamura Y, Alexeev V, Mastrangelo MJ, Selvan SR (2011) Interleukin 10 in the tumor

microenvironment: a target for anticancer immuno-therapy. Immunol Res 51(2–3):170–182

41. Hart KM, Byrne KT, Molloy MJ, Usherwood EM, Berwin B (2011) IL-10 immunomodulation of myeloid cells regulates a murine model of ovarian cancer. Front Immunol 2:29

42. Zhang L, Liu W, Wang X, Wang X, Sun H (2019) Prognostic value of serum IL-8 and IL-10 in patients with ovarian cancer undergoing chemotherapy. Oncol Lett 17(2):2365–2369

43. Li FZ, Dhillon AS, Anderson RL, McArthur G, Ferrao PT (2015) Phenotype switching in mela-noma: implications for progression and therapy. Front Oncol 5:31

44. Park GB, Chung YH, Kim D (2017) Induction of galectin-1 by TLR-dependent PI3K activa-tion enhances epithelial-mesenchymal transi-tion of metastatic ovarian cancer cells. Oncol Rep 37(5):3137–3145

45. Liu CY, Xu JY, Shi XY et al (2013) M2-polarized tumor-associated macrophages promoted epithelial-mesenchymal transition in pancreatic cancer cells, partially through TLR4/IL-10 signaling pathway. Lab Investig 93(7):844–854

46. Gemelli C, Marani TZ, Bicciato S et al (2014) MafB is a downstream target of the IL-10/STAT3 signaling pathway, involved in the regulation of macrophage de-activation. Biochimi Biophys Acta 1843(5):955–964

47. Liu Y, Gong W, Yang ZY et al (2017) Quercetin induces protective autophagy and apoptosis through ER stress via the p-STAT3/Bcl-2 axis in ovarian can-cer. Apoptosis 22(4):544–557

48. Lo H-W, Hsu S-C, Xia W et al (2007) Epidermal growth factor receptor cooperates with signal trans-ducer and activator of transcription 3 to induce epithelial-mesenchymal transition in cancer cells via up-regulation of TWIST gene expression. Cancer Res 67(19):9066–9076

49. Santoiemma PP, Powell DJ Jr (2015) Tumor infiltrat-ing lymphocytes in ovarian cancer. Cancer Biol Ther 16(6):807–820

50. Ghisoni E, Imbimbo M, Zimmermann S, Valabrega G (2019) Ovarian cancer immunotherapy: turning up the heat. Int J Mol Sci 20(12):2927

51. Grivennikov SI, Greten FR, Karin M (2010) Immunity, inflammation, and cancer. Cell 140(6):883–899

52. Steinman RM, Banchereau J (2007) Taking dendritic cells into medicine. Nature 449(7161):419–426

53. Chen YL, Chang MC, Chen CA, Lin HW, Cheng WF, Chien CL (2012) Depletion of regulatory T lymphocytes reverses the imbalance between pro- and anti-tumor immunities via enhancing anti-gen-specific T cell immune responses. PLoS One 7(10):e47190

54. Chae CS, Teran-Cabanillas E, Cubillos-Ruiz JR (2017) Dendritic cell rehab: new strategies to unleash therapeutic immunity in ovarian cancer. Cancer Immunol Immunother 66(8):969–977

55. Tanyi JL, Chu CS (2012) Dendritic cell-based tumor vaccinations in epithelial ovarian cancer: a system-atic review. Immunotherapy 4(10):995–1009

56. Wertel I, Polak G, Tarkowski R, Kotarska M (2011) Evaluation of IL-10 and TGF-beta levels and myeloid and lymphoid dendritic cells in ovarian can-cer patients. Ginekol Pol 82(6):414–420

57. Liu WH, Liu JJ, Wu J et al (2013) Novel mechanism of inhibition of dendritic cells maturation by mesen-chymal stem cells via interleukin-10 and the JAK1/STAT3 signaling pathway. PLoS One 8(1):e55487

58. Posselt G, Schwarz H, Duschl A, Horejs-Hoeck J (2011) Suppressor of cytokine signaling 2 is a feedback inhibitor of TLR-induced activation in human monocyte-derived dendritic cells. J Immunol 187(6):2875–2884

59. Conzelmann M, Wagner AH, Hildebrandt A et al (2010) IFN-gamma activated JAK1 shifts CD40-induced cytokine profiles in human antigen-presenting cells toward high IL-12p70 and low IL-10 production. Biochem Pharmacol 80(12):2074–2086

60. Sipak-Szmigiel O, Wlodarski P, Ronin-Walknowska E et al (2017) Serum and peritoneal fluid concen-trations of soluble human leukocyte antigen, tumor necrosis factor alpha and interleukin 10 in patients with selected ovarian pathologies. J Ovarian Res 10(1):25

61. Liu CZ, Zhang L, Chang XH et al (2012) Overexpression and immunosuppressive functions of transforming growth factor 1, vascular endothelial growth factor and interleukin-10 in epithelial ovar-ian cancer. Chin J Cancer Res 24(2):130–137

62. Zhang S-N, Choi I-K, Huang J-H, Yoo J-Y, Choi K-J, Yun C-O (2011) Optimizing DC vaccination by combination with oncolytic adenovirus coexpressing IL-12 and GM-CSF. Mol Ther 19(8):1558–1568

63. Wallet MA, Sen P, Tisch R (2005) Immunoregulation of dendritic cells. Clin Med Res 3(3):166–175

64. Maccalli C, Parmiani G, Ferrone S (2017) Immunomodulating and immunoresistance prop-erties of cancer-initiating cells: implications for the clinical success of immunotherapy. Immunol Investig 46(3):221–238

65. Batchu RB, Gruzdyn OV, Qazi AM et al (2016) Pancreatic cancer cell lysis by cell-penetrating peptide-MAGE-A3-induced cytotoxic T lympho-cytes. JAMA Surg 151(11):1086–1088

66. Batchu RB, Gruzdyn OV, Qazi A, Mahmud EM, Weaver DW, Gruber SA (2016) Pancreatic cancer-induced microenvironment inhibits dendritic cell activation via decreased nuclear localization of NF-kB. 73rd central surgical association, Montreal, Quebec, 10–12 Mar 2016

67. Batchu RB, Qazi A, Gruzdyn OV et al (2015) Inhibition of epithelial ovarian cancer (EOC)-induced microenvironment can restore dendritic cell (DC) activation and migration. J Am Coll Surg 221(4):e33

68. Batchu RB, Qazi A, Gruzdyn OV et al IL-10 and TGF-b blockade reverses the inhibitory effects of pancreatic carcinoma on dendritic cell activation and

migration. Paper presented at: 39th association of VA surgeons, Miami Beach, FL, 2–5 May 2015

69. Batchu RB, Gruzdyn OV, Moreno-Bost AM et al (2014) Efficient lysis of epithelial ovarian cancer cells by MAGE-A3-induced cytotoxic T lymphocytes using rAAV-6 capsid mutant vector. Vaccine 32(8):938–943

70. Batchu RB, Gruzdyn OV, Kung ST, Gruber SA, Weaver DW (2014) Dendritic cell based immunotherapy of cancer with cell penetrating domains. Indian J Surg Oncol 5(1):3–4

71. Batchu RB, Gruzdyn O, Potti RB, Weaver DW, Gruber SA (2014) MAGE-A3 with cell-penetrating domain as an efficient therapeutic cancer vaccine. JAMA Surg 149(5):451–457

72. Batchu RB, Gruzdyn OV, Kolli BK et al (2020) Inhibition of IL-10 in the tumor microenvironment potentiates mesothelin-chimeric antigen receptor NK-92MI-mediated killing of ovarian cancer cells. J Am Coll Surg

73. Batchu RB, Gruzdyn OV, Tavva PS et al (2019) Engraftment of mesothelin chimeric antigen receptor using a hybrid sleeping beauty/minicircle vector into NK-92MI cells for treatment of pancreatic cancer. Surgery 166(4):503–508

74. Batchu RB, Gruzdyn OV, Rohondia S et al (2018) Inhibition of IL-10 augments mesothelin chimeric antigen receptor T cell activity in epithelial ovarian cancer. J Am Coll Surg 227(4):e215–e216

75. Batchu RB, Gruzdyn OV, Mahmud EM et al (2018) Inhibition of Interleukin-10 in the tumor microenvironment can restore mesothelin chimeric antigen receptor T cell activity in pancreatic cancer in vitro. Surgery 163(3):627–632

76. Gruzdyn O, Batchu RB, Mahmud EM et al (2017) Mesothelin chimeric antigen receptor (CAR)-mediated therapy for ovarian cancer. J Am Coll Surg 225(4):e47

77. Carr TM, Adair SJ, Fink MJ, Hogan KT (2008) Immunological profiling of a panel of human ovarian cancer cell lines. Cancer Immunol Immunother 57(1):31–42

78. Zhang S, Zhou X, Yu H, Yu Y (2010) Expression of tumor-specific antigen MAGE, GAGE and BAGE in ovarian cancer tissues and cell lines. BMC Cancer 10:163

79. Coolen AL, Lacroix C, Mercier-Gouy P et al (2019) Poly(lactic acid) nanoparticles and cell-penetrating peptide potentiate mRNA-based vaccine expression in dendritic cells triggering their activation. Biomaterials 195:23–37

80. Lim S, Koo JH, Choi JM (2016) Use of cell-penetrating peptides in dendritic cell-based vaccination. Immune Netw 16(1):33–43

81. Chauhan A, Tikoo A, Kapur AK, Singh M (2007) The taming of the cell penetrating domain of the HIV tat: myths and realities. J Control Release 117(2):148–162

82. Batchu RB, Moreno AM, Szmania SM et al (2005) Protein transduction of dendritic cells for NY-ESO-1-based immunotherapy of myeloma. Cancer Res 65(21):10041–10049

83. Brichard VG, Lejeune D (2007) GSK's antigen-specific cancer immunotherapy programme: pilot results leading to phase III clinical development. Vaccine 25(Suppl 2):B61–B71

84. Schultz ES, Lethe B, Cambiaso CL et al (2000) A MAGE-A3 peptide presented by HLA-DP4 is recognized on tumor cells by CD4+ cytolytic T lymphocytes. Cancer Res 60(22):6272–6275

85. Daya S, Berns KI (2008) Gene therapy using adeno-associated virus vectors. Clin Microbiol Rev 21(4):583–593

86. Ponnazhagan S, Mahendra G, Curiel DT, Shaw DR (2001) Adeno-associated virus type 2-mediated transduction of human monocyte-derived dendritic cells: implications for ex vivo immunotherapy. J Virol 75(19):9493–9501

87. Korokhov N, de Gruijl TD, Aldrich WA et al (2005) High efficiency transduction of dendritic cells by adenoviral vectors targeted to DC-SIGN. Cancer Biol Ther 4(3):289–294

88. Ussher JE, Taylor JA (2010) Optimized transduction of human monocyte-derived dendritic cells by recombinant adeno-associated virus serotype 6. Hum Gene Ther 21(12):1675–1686

89. Vazquez J, Chavarria M, Lopez GE et al (2020) Identification of unique clusters of T, dendritic and innate lymphoid cells in the peritoneal fluid of ovarian cancer patients. Am J Reprod Immunol 84(3):e13284

90. Brencicova E, Jagger AL, Evans HG et al (2017) Interleukin-10 and prostaglandin E2 have complementary but distinct suppressive effects on toll-like receptor-mediated dendritic cell activation in ovarian carcinoma. PLoS One 12(4):e0175712

91. Goyne HE, Stone PJ, Burnett AF, Cannon MJ (2014) Ovarian tumor ascites CD14+ cells suppress dendritic cell-activated CD4+ T-cell responses through IL-10 secretion and indole amine 2,3-dioxygenase. J Immunother 37(3):163–169

92. Scholz C, Rampf E, Toth B et al (2009) Ovarian cancer-derived glycodelin impairs in vitro dendritic cell maturation. J Immunother 32(5):492–497

93. Zavadova E, Savary CA, Templin S, Verschraegen CF, Freedman RS (2001) Maturation of dendritic cells from ovarian cancer patients. Cancer Chemother Pharmacol 48(4):289–296

94. Melichar B, Touskova M, Tosner J, Kopecky O (2001) The phenotype of ascitic fluid lymphocytes in patients with ovarian carcinoma and other primaries. Onkologie 24(2):156–160

95. Wang Y, Yi J, Chen X, Zhang Y, Xu M, Yang Z (2016) The regulation of cancer cell migration by lung cancer cell-derived exosomes through TGF-beta and IL-10. Oncol Lett 11(2):1527–1530

96. Huang X, Zou Y, Lian L et al (2013) Changes of T cells and cytokines TGF-beta1 and IL-10 in mice during liver metastasis of colon carcinoma: implications for liver anti-tumor immunity. J Gastrointest Surg 17(7):1283–1291

97. Spinardi-Barbisan AL, Barbisan LF, de Camargo JL, Rodrigues MA (2004) Infiltrating γ-δ T lymphocytes, natural killer cells, and expression of IL-10 and TGF-beta1 in chemically induced neoplasms in male Wistar rats. Toxicol Pathol 32(5):548–557

98. Hirao M, Onai N, Hiroishi K et al (2000) CC chemokine receptor-7 on dendritic cells is induced after interaction with apoptotic tumor cells: critical role in migration from the tumor site to draining lymph nodes. Cancer Res 60(8):2209–2217

99. Montfort A, Pearce O, Maniati E et al (2017) A strong B-cell response is part of the immune landscape in human high-grade serous ovarian metastases. Clin Cancer Res 23(1):250–262

100. Nelson BH (2008) The impact of T-cell immunity on ovarian cancer outcomes. Immunol Rev 222:101–116

101. Moore KW, de Waal MR, Coffman RL, O'Garra A (2001) Interleukin-10 and the interleukin-10 receptor. Annu Rev Immunol 19(1):683–765

102. Boks MA, Kager-Groenland JR, Haasjes MS, Zwaginga JJ, van Ham SM, ten Brinke A (2012) IL-10-generated tolerogenic dendritic cells are optimal for functional regulatory T cell induction—a comparative study of human clinical-applicable DC. Clin Immunol 142(3):332–342

103. Kobie JJ, Wu RS, Kurt RA et al (2003) Transforming growth factor β inhibits the antigen-presenting functions and antitumor activity of dendritic cell vaccines. Cancer Res 63(8):1860–1864

104. Liou JT, Mao CC, Ching-Wah Sum D et al (2013) Peritoneal administration of Met-RANTES attenuates inflammatory and nociceptive responses in a murine neuropathic pain model. J Pain 14(1):24–35

105. Prazeres PHDM, Leonel C, Walison NS et al (2020) Ablation of sensory nerves favours melanoma progression. J Cell Mol Med 24(17):9574–9589

106. Turnquist HR, Lin JT, Ashour AE et al (2007) IL-12 induces extensive intratumoral immune cell infiltration and specific anti-tumor cellular immunity. Int J Oncol 30(3):631–639

107. Thepmalee C, Panya A, Junking M, Chieochansin T, Yenchitsomanus PT (2018) Inhibition of IL-10 and TGF-beta receptors on dendritic cells enhances activation of effector T-cells to kill cholangiocarcinoma cells. Hum Vaccin Immunother 14(6):1423–1431

108. Gordy JT, Luo K, Francica B, Drake C, Markham RB (2018) Anti-IL-10-mediated enhancement of antitumor efficacy of a dendritic cell-targeting MIP3α-gp100 vaccine in the B16F10 mouse melanoma model is dependent on type I interferons. J Immunother 41(4):181–189

109. Yang L, Dong Y, Li Y et al (2019) IL-10 derived from M2 macrophage promotes cancer stemness via JAK1/STAT1/NF-kappaB/Notch1 pathway in non-small cell lung cancer. Int J Cancer 145(4):1099–1110

110. Harrington BS, Annunziata CM (2019) NF-kappaB signaling in ovarian cancer. Cancers (Basel) 11(8):1182

111. Ignacio RM, Kabir SM, Lee ES, Adunyah SE, Son DS (2016) NF-kappaB-mediated CCL20 reigns dominantly in CXCR2-driven ovarian Cancer progression. PLoS One 11(10):e0164189

112. Yang N, Huang J, Greshock J et al (2008) Transcriptional regulation of PIK3CA oncogene by NF-kappaB in ovarian cancer microenvironment. PLoS One 3(3):e1758

Targeted Delivery of IL-12 Adjuvants Immunotherapy by Oncolytic Viruses

4

Andrea Vannini, Valerio Leoni,
and Gabriella Campadelli-Fiume

Abstract

The great hopes raised by the discovery of the immunoregulatory cytokine interleukin 12 (IL-12) as an anticancer agent were marred during early clinical experimentation because of severe adverse effects, which prompted a search for alternative formulations and routes of administration. Onco-immunotherapeutic viruses (OIVs) are wild-type or genetically engineered viruses that exert antitumor activity by causing death of the tumor cells they infect and by overcoming a variety of immunosuppressive mechanisms put in place by the tumors. OIVs have renewed the interest in IL-12, as they offer the opportunity to encode the cytokine transgenically from the viral genome and to produce it at high concentrations in the tumor bed. A large body of evidence indicates that IL-12 serves as a potent adjuvant for the immunotherapeutic response elicited by OIVs in murine tumor models. The list of OIVs includes onco-immunotherapeutic herpes simplex, adeno, measles, Newcastle disease, and Maraba viruses, among others.

The large increase in IL-12-mediated adjuvanticity was invariably observed for all the OIVs analyzed. Indirect evidence suggests that locally delivered IL-12 may also increase tumor antigenicity. Importantly, the OIV/IL-12 treatment was not accompanied by adverse effects and elicited a long-lasting immune response capable of halting the growth of distant tumors. Thus, OIVs provide an avenue for reducing the clinical toxicity associated with systemic IL-12 therapy, by concentrating the cytokine at the site of disease. The changes to the tumor microenvironment induced by the IL-12-armed OIVs primed the tumors to an improved response to the checkpoint blockade therapy, suggesting that the triple combination is worth pursuing in the future. The highly encouraging results in preclinical models have prompted translation to the clinic. How well the IL-12–OIV–checkpoint inhibitors' combination will perform in humans remains to be fully investigated.

Keywords

IL-12 · Oncolytic virus · Onco-immunotherapeutic virus · Oncolytic immunotherapy · Cancer therapy · OIV · HSV · Adenovirus · Maraba · Measles · NDV · Immune heating · TME · Proinflammatory

A. Vannini · V. Leoni · G. Campadelli-Fiume (✉)
Department of Experimental, Diagnostic and Specialty Medicine, University of Bologna, Bologna, Italy
e-mail: andrea.vannini5@unibo.it;
gabriella.campadelli@unibo.it

67

A. Birbrair (ed.), *Tumor Microenvironment*, Advances in Experimental Medicine and Biology 1290,
https://doi.org/10.1007/978-3-030-55617-4_4

cytokines · Adjuvanticity · Antigenicity · CP blockade · CPI combination

4.1   IL-12, Basic Features

Interleukin 12 (IL-12) is a proinflammatory and immunoregulatory cytokine discovered independently by two laboratories 30 years ago [1–3]. Structurally, IL-12 (p70) is a disulfide-linked heterodimer of IL-12A (p35) and IL-12B (p40) subunits, shared with two additional members of the IL-12 family, IL-35 and IL-23, respectively [4]. IL-12 is expressed and secreted by activated macrophages, dendritic cells (DCs), microglia, monocytes, neutrophils, and B cells in response to microbial infections and malignancies [5]. IL-12 binds the IL-12 receptor (IL-12R), a beta1-beta2 heterodimer type I cytokine receptor expressed by a number of cells, including natural killer (NK), activated T, and natural killer T (NKT) cells, DCs, B cells, and macrophages [6–9]. Each of these cells responds to IL-12 through specific signaling pathways and responses. Shortly after the IL-12 discovery, it was recognized that IL-12 exerts a strong adjuvant effect with antipathogens [10] and anticancer vaccines [11, 12]. This cytokine promotes the secretion of interferon gamma (IFNγ) by NK, T, and B cells [13] and additional proinflammatory cytokines, including tumor necrosis factor alpha (TNFα) and granulocyte-macrophage colony-stimulating factor (GM-CSF). In turn, these molecules target, recruit, and activate effector cells of the innate immune response, and, together, they make IL-12 a master regulator. Importantly, IL-12 provides a link between the activation of innate and adaptive responses by priming Th1 cells for activation. The latter is a key part of the anticancer response, as it promotes the reactivation of memory CD4+ T cells and their repolarization from tumor-permissive Th2 to antitumor Th1 cells [14]. IL-12 also triggers NK and CD8+ T-cell activation, proliferation, and differentiation [15], leading to the generation of cytotoxic T cells (CTLs). Specifically, the cytokine primes macrophages for antigen presentation [8] and their M2 to M1

repolarization [16] promotes DC maturation and activity [17] and induces B-cell proliferation, differentiation, and an IgE to IgG1 shift [6].

IL-12 also affects the nonimmune cells of the tumor microenvironment (TME), including stromal cells and blood vessels that feed the tumor and sustain carcinogenesis [18]. Mechanistically, IL-12 downregulates the proangiogenic cytokines CCL2, CCL6, IL-6, VEGF, and other factors, and upregulates angiostatic and antiangiogenic factors, including TNFα, IFNα, IFNβ, IFNγ, CXCL9, and CXCL10 [19, 20]. Finally, the cytokine facilitates immune cells' recruitment and lymphocyte localization to the tumor through IFNγ-dependent cascades and upregulation of immune-attractants [21]. Globally, IL-12 reprograms the tumor TME from a protumoral hospitable alcove to an antitumor environment.

The potent anticancer effects elicited by systemically administered IL-12 were well documented in preclinical models (reviewed in [22]). However, early studies in humans were marred by limited efficacy and generalized toxicity. The severe to lethal effects included hematopoietic suppression and gastrointestinal, muscular, pulmonary and liver toxicity, and dysfunction [23–26]. These side effects prompted the search for novel formulations and for new administration strategies capable of achieving higher local IL-12 concentrations. A promising approach consisted in intratumoral delivery by nonreplicating adenoviruses (AdVs) [27, 28]. Additional approaches include subcutaneous injections of the recombinant protein, fine-tunable expression systems [29], delivery of IL-12-encoding plasmid in the tumor bed, coupling of the cytokine with a tumor-targeting antibody (Ab) [30, 31], and transgenic expression by engineered tumor-specific CAR-T, autologous immune or cancer cells [32, 33]. The Clinicaltrials.gov website lists 84 active clinical trials for the evaluation of IL-12 treatments for a variety of solid tumors, including pancreatic, prostatic, colorectal, ovarian, breast, and liver cancers. IL-12 is administered as a recombinant protein (38), as a fusion protein with cancer-specific antibodies (2), is expressed by a plasmid (23), is vectored by bacteria (1), by viruses (15), or by cells, including CAR-Ts (2), or engineered autologous cells (3).

4.2   From Oncolytic to Onco-Immunotherapeutic Viruses, a Paradigm Shift

Oncolytic viruses (OVs) are replication-competent wild-type (wt) or engineered viruses that selectively replicate in tumor cells and/or in cells in the TME. The intrinsic properties of tumors are immune tolerance and immune evasion, which, together with defects in innate immune responses, greatly favor virus susceptibility and replication. Early preclinical studies on human cancer cells implanted in nude mice highlighted the antitumor efficacy exerted by OVs, mainly as a consequence of lysis of the infected cells by immunogenic cell death mechanisms, including necroptosis [34]. When preclinical models were shifted to immunocompetent mouse models, it became apparent that, in addition to tumor cell lysis, tumor infection by oncolytic viruses resulted in the tolerance breakdown, the induction of an innate response to the tumor and, ultimately, to immune control of tumor growth. The current resurgence of interest in OVs is the result of an array of effects, among which are the secretion of type I and II interferons and other proinflammatory cytokines, the infiltration of tumors by NK cells and T-lymphocytes, the activation of these cells, and an overall reprogramming of the TME that enhances the adaptive systemic antitumor response. In brief, OVs convert immunologically cold tumors into immunologically hot ones [35]. Through these modifications, OV-infected tumor cells serve as antigen agnostic antitumor vaccines [36–38]. These OVs can be renamed as onco-immunotherapeutic viruses (OIVs).

The OIV-mediated immunotherapeutic effects observed in preclinical models were documented in humans, in particular with talimogene laherparepvec (T-Vec), a mildly attenuated oHSV that expresses GM-CSF to increase macrophage, DC, and neutrophil responses. In cutaneous melanoma patients, the intratumoral administration of T-Vec in some of the lesions resulted in the shrinkage of distant untreated lesions, even though the reduction was not as large as that observed in the treated lesions [39]. The distant response is attributed to an abscopal immune effect, caused by the adaptive immune response to the tumor.

The immunotherapy of cancer has been recently revolutionized by checkpoint inhibitors (CPIs). Unfortunately, their activity is exerted only toward a subset of cancers and to a fraction of patients, and is limited by severe adverse effects. Making tumors immunologically hot by OIVs confers CPI susceptibility to tumors that are otherwise resistant [40]. Thus, OIVs represent the ideal partners for checkpoint blockade [40–43]. Today, OIVs are considered as most promising tools to increase the efficacy and broaden the spectrum of CPIs.

The ability of OIVs to unleash immune suppression and to elicit an innate response to tumors, even in highly immune suppressive tumors, renders OIVs the ideal companions for IL-12. Furthermore, the IL-12 gene can be expressed transgenically from the viral genome in the tumor bed, so as to prevent systemic toxicity of the cytokine. By exerting its adjuvant effect, IL-12 promotes the shift from an innate response to the virus toward an adaptive long-term memory response to the tumor [44, 45] (Fig. 4.1).

The list of IL-12-expressing OIVs and the beneficial effects of OIV-delivered IL-12 in the tumor bed has been documented by numerous studies. Here, we review some select examples.

4.3   oHSVs

In a highly innovative, seminal study, Martuza, Rabkin, and coworkers recognized the ability of oHSVs to confer protection not only in the treated tumor through lysis of the infected cancer cells—the dominant paradigm at that time—but also through elicitation of the host immune response [46]. They employed the oHSV named G207 as a helper virus to generate dvIL12/G207, a defective HSV vector expressing IL-12 [46].

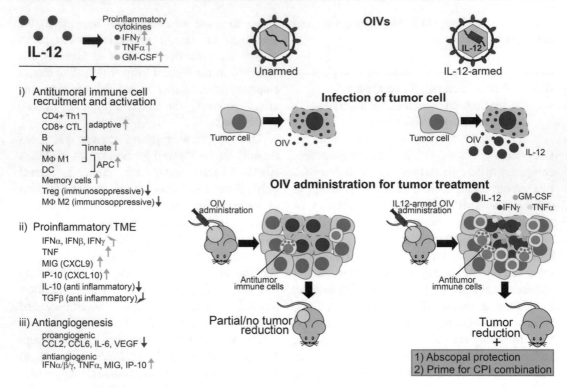

Fig. 4.1 Schematic summary of how IL-12 contributes to antitumor immunity and main effects elicited by IL-12-expressing OIVs as compared with those elicited by unarmed OIVs. The IL-12 cassette is engineered in the OIV genome so that IL-12 is transgenically expressed by the OIVs in the infected cells, i.e., the cancerous cells and – in some cases – cells of the TME. Infection provokes cell death. When present, IL-12 induces expression of pro-inflammatory cytokines, IFNγ, TNFα, and GM-CSF, among others. These (1) recruit antitumoral immune cells to the tumor and activate them, (2) reshape the immunosuppressive TME to a proinflammatory and anticancer setting, (3) contrast tumor angiogenesis by reducing proangiogenic and increasing antiangiogenic factors. As a consequent of the combined effects of OIVs and the IL-12-induced immune modifications to the TME, treatment with IL-12-expressing OIVs results in higher inhibition of tumor growth relative to treatment with unarmed versions of OIVs. The administration of IL-12 expressing OIVs also results in higher abscopal protection than the administration of unarmed OVs, and primes for the CP blockade

G207 and its derivatives harbor deletions in UL39 (ICP6) and both copies of the $\gamma_1 34.5$ gene; hence, they are highly safe yet attenuated. Remarkably, the intralesionally treated tumors responded to the therapy and exhibited reduced growth; 33% of the mice were tumor free (TF). Moreover, the treatment reduced the growth of untreated contralateral tumors; i.e., elicited a long-term in situ vaccination effect [46] dependent on the systemic T-cell response. To improve the antigen presentation, the IL-12 gene was engineered in the G207-derived G47Δ backbone, which additionally harbors the US12 (α47) deletion (G47Δ-mIL12 virus) [47]. This oHSV was used against murine glioblastoma and

showed T-cell-dependent reduced tumor growth, reduced intratumoral Treg levels, and inhibition of angiogenesis [47].

Fong and coworkers focused on the development of an IL-12 oHSV for the treatment of squamous cell carcinoma and colorectal and liver cancers [48–50]. The NV1042 viral genome carried multiple deletions, namely, US10, US11, and α47 genes and one copy of the $\gamma_1 34.5$, α0, α4 genes. Initially, the authors compared the effects of the transgenic expression of IL-12 to that of GM-CSF and found that the IL-12 virus was superior in various models [50–52]. The production of the α-promoter-driven IL-12 ranged from 1 to 35 ng/mL at 24–72 h post infection per

5×10^4 cells infected at one plaque-forming unit (PFU) per cell [48]. Although the virus was overall attenuated, a single injection reduced the growth of CT26 murine colorectal cancers; a few mice were completely cured [49]. The local immuno stimulation provided by IL-12 resulted in the control of hepatic challenge tumors upon resection of the primary tumor [49]. In a pulmonary metastatic model, the immunotherapeutic effect of IL-12 virus was shown to involve CD4+ and CD8+ T cells [53, 54]. In a model of squamous cell carcinoma, it was also verified that treatment with IL-12-expressing NV1042 resulted in anti-angiogenic effects [53, 54]. The same virus also proved effective upon systemic administration against the pulmonary metastases of squamous cell carcinoma [53, 54], liver metastases of colorectal cancer [55], and of poorly immunogenic prostate adenocarcinoma and metastatic prostatic cancers [51, 52, 56].

Markert and coworkers engineered the IL-12 gene in a less attenuated oHSV, named M002, initially designed for glioblastoma treatment. M002 carries the replacement of both copies of the $\gamma_1 34.5$ gene with IL-12 and no other virus genome modification. The two copies of mIL-12 were placed under the murine egr-1 promoter. In vitro, the extent of IL-12 expression was in the range of 0.8–3.2 ng/mL per 5×10^5 cells infected at 1 PFU/cell, at 24 h after infection [57]. Importantly, the M002 treatment resulted in a significant increase in mouse survival [57]. The same virus was also effective in preclinical models of breast cancer metastases to the brain [58], glioma [59], neuroblastoma [60], rhabdomyosarcoma [61], undifferentiated sarcoma [62], and in pediatric high-grade brain tumor and medulloblastoma xenografts [63, 64]. M002 and its cognate M032, an identical recombinant virus expressing hIL-12 in place of mIL-12, have undergone detailed safety analyses in mice and in *Aotus* primates [65, 66]. A phase 1 clinical trial for glioblastoma multiforme treatment is recruiting participants [67].

Altogether, the three series of studies highlight the superior effects of IL-12-armed oHSVs against murine primary tumors as well as distant T-cell-based immune protection.

A general notion that permeates the OIV field, and particularly the oHSV field, is that the initial safety concerns led to viruses that exhibited a very high safety profile in mice as well as in humans and are effective against murine tumor models but not as effective against human tumors. These considerations led to calls for "safe-and-robust" oHSVs that are more effective than those that are now in clinical practice or trials. To this end, our laboratory engineered tropism-retargeted oHSVs whose safety rests on cancer-specific tropisms, rather than on the cancer-selective replication typical of the oHSVs that are attenuated to varying degrees. The principle of tropism retargeting hinges on two series of modifications, namely, the ablation of HSV tropism for the natural receptors nectin1 and herpesvirus entry mediator (HVEM), and the readdress of the tropism (retargeting) to a cancer-specific receptor of choice. Retargeting was obtained by engineering a single chain Ab (or a ligand) in the receptor-binding glycoprotein, gD [68–70]. The cancer receptor we have chosen is the human epithelial growth factor receptor 2 (HER2), expressed in a subset of breast, gastric, gastroesophageal, lung, and other types of cancers [71, 72]. The HER2-retargeted R-LM113 recombinant virus was then armed with IL-12 to generate R-115 [68, 73]. These viruses carry no deletion or mutation in any gene other than the glycoprotein D gene; hence, they are "fully virulent" to their targeted HER2-positive cancer cells. A direct comparison showed that the IL-12-expressing R-115 virus was more efficacious against Lewis lung carcinoma 1 cells expressing human HER2 (LLC-1-HER2) than was the unarmed R-LM113 and conferred distant long-lasting immune protection against challenge tumors [73]. To obtain mechanistic insight into how IL-12 expressed in the tumor bed contributed to the immune therapeutic effect, we compared the tumor-infiltrating lymphocytes and key immune proteins in tumors from mice treated with IL-12-armed or unarmed retargeted oHSVs. In the R-115-treated group, the tumors from responder mice were infiltrated with CD4+ and CD8+ T cells and their activated CD69+ subpopulation, with CD335+ NK cells and their activated CD69+ subpopulation, and with CD141+ DC cells. The tumors were additionally characterized by a decrease in CD11b+ monocytes/macrophages and an increase in the proinflammatory factors IFNγ, IL-2, TNFα, and t-bet [73]. Intriguingly, in the same tumors (from R-115-treated responder mice), there was an increase in anti-inflammatory factors, such as Tregs, tumor PD-L1, and IL-10. The R-LM113-treated mice which underwent tumor reduction recapitulated these responses but to a lesser extent. Altogether, in the LLC-1 model, the immune heating of the tumors and the simultaneous increase in immune-suppressive factors were primed for check point blockade. Thus, the co-administration of R-115 with anti-PD-1 increased the proportion of cured mice from 20 (virus alone) to 60%. The further addition of anti-CTLA-4 cured 100% of the mice (our unpublished results). A notable effect of the treatment with the IL-12-expressing R-115 was an increase in the reactivity of splenocytes and antibodies to tumor cells. The increase was even higher in the mice treated with the combination of R-115 and anti-PD-1. The results suggest that IL-12 boosted not only the adaptive response but also augmented the repertoire of T and B cells that were reactive to tumor-specific antigens. They raise the possibility that IL-12 expression in the tumor bed also increased tumor antigenicity.

These findings were confirmed and extended in a highly immunosuppressive, transplantable glioblastoma model that recapitulated human glioblastoma [74]. A single orthotopic injection of 10^6 PFU of the IL-12-armed R-115 administered to well-established tumors immediately before the appearance of symptoms caused tumor regression and spared approximately 25% of the mice. The tumor specimens showed CD4+ and CD8+ lymphocytes deeply infiltrating into tumor masses [74]. Thus, HER2-retargeted fully virulent oHSVs emerge as professional igniters of antitumor immunity and IL-12 was their optimal partner.

To summarize, among all the oHSVs analyzed—whether attenuated to varying degrees or not—and in any tumor model tested, IL-12 greatly augmented the OIV-mediated protection against the primary tumor, favored conspicuous

immune modifications to the immunosuppressive microenvironment, and contributed to distant abscopal protection; i.e., they had an antigen agnostic vaccination effect. These effects can be interpreted as the result of increased adjuvanticity and, possibly, of increased antigenicity.

4.4 Adenoviruses

First-generation oncolytic adenoviruses (oAdV) consisted of replication-incompetent viruses, the cancer selectivity of which depended on attenuation provided by knocking out the E1 and E3 genes. E1 is an essential gene; its deletion allows for only one round of replication and prevents uncontrolled virus replication in host tissues. Deletion of the nonessential E3 gene abrogates the major immune-escape mechanisms; the virus becomes unable to counteract the antiviral responses of the infected cells, and its replication is restricted to tumor cells defective in the innate response [75]. The insertion of transgenes in the viral backbone enabled their expression in the infected cancer cells and restricted their accumulation to the tumor bed. Among the engineered cargos were genes encoding cytokines, antigens, tumor suppressors, and suicide proteins [76]. In a comparative study with IL-2 and HSV-1 thymidine kinase (TK), IL-12 emerged clearly as a superior payload [27]. Specifically, a single intratumoral injection of the IL-12-encoding AdV into liver metastatic colon carcinoma and breast cancer cells significantly reduced tumor growth and improved mouse survival [27, 77]. The unarmed virus showed no therapeutic effect. The enhanced efficacy conferred by IL-12 was confirmed in a number of tumor models [78]. In mice bearing CT-26 colon carcinomas, tumor growth was reduced and most of the mice were tumor-free after a single intratumoral treatment. Importantly, a systemic response was elicited, measured as protection from distant untreated tumors and reactive antitumor lymphocytes [79]. When employed against poorly immunogenic tumors, i.e., glioma, prostatic, and thyroid cancers, AdV-IL-12 caused a significant reduction in primary tumor growth; some mice were completely cured; CD4+ and CD8+ T-cell infiltrated tumor masses, and long-lasting protection was established in an IL-12-dependent fashion [80–83].

Second generation oAdVs consisted of replication-competent viruses that harbored smaller deletions in the E1 and E3 genes; hence, they were still partially attenuated. [84–86]. Because the viral load in the tumor bed increased over time, the level of IL-12 also increased. In vitro, the concentrations were on the order of 4–10 μg/10^6 cells 48 h after infection, i.e., 80–200-fold higher than those observed with replication-deficient AdV-IL-12 vectors [81, 87]. Replication-competent AdVs proved highly effective as antitumor agents and completely protected 50% of the mice [84]. The reduction in efficiency upon immune cell depletion implied that CD4+/CD8+ T cells [79, 88] and NK cells [81, 87, 89] were the immune populations which contributed more to the anticancer response.

These encouraging results prompted the search for numerous improvements. Thus, for safety and efficacy purposes, a tunable form of IL-12 was obtained by placing the IL-12 gene under a conditionally activated promoter (Ad-RTS-IL-12) [29, 90, 91]. In an alternative approach, IL-12 was engineered as a nonsecreted form [92] or as a p35-p40 fusion protein, named FIL-12 [87, 93]. Both modifications resulted in higher therapeutic activity. Another approach was based on the notion that IL-12 synergizes with a variety of antitumor factors. AdV vectors for combinatorial expression included the proinflammatory factors IL-23, IL-18, GM-CSF, CD80, and 4-1BBL [86, 88, 94–99]; anti-immunosuppressive factors, such as shVEGF, decorin, and anti-PD-L1 [100–103]; and suicidal genes, such as cytosine deaminase (yCD) and HSV-1 thymidine kinase (TK) [87]. An example is Ad5-yCD/mutTKSR39rep-hIL12, which encodes IL-12, yCD, and a mutant form of TK. The enzymes convert systemically administered prodrugs to their active forms, which in turn inhibit DNA synthesis in infected cells. When administered to mice bearing TRAMP-C2 prostate adenocarcinoma as monotherapy, the virus elicited NK and T-cell responses, and cured 40%

of the mice. In combination with the yCD- and TK-activated prodrugs the virus cured 70–80% of the mice [87]. In preclinical studies, replication-competent AdV coexpressing IL-12/ IL-18, IL-12/IL-23, or IL-12/4-1BBL caused a complete response in mice harboring poorly immunogenic B16-F10 melanoma [86, 94, 95]. In the same model, the combination of IL-12 with GM-CSF or shVEGF led to complete response in 90 and 60% of the mice, respectively [88, 100]. The IL-12/decorin combination proved effective in weakly immunogenic 4 T1 tumors that are refractory to IL-12 as a consequence of high intratumoral TGF-β levels and Treg infiltration [101].

In other developments, oAdVs were administered together with therapies such as radiation, DC infusion, and CAR-T [96, 99, 102, 103]. All approaches resulted in effective therapeutic responses. In head and neck squamous cell carcinoma (HNSCC), local treatment with AdV coexpressing IL-12 and anti-PD-L1 was primed for systemic CAR-T-cell therapy and significantly improved mouse survival [102, 103]. Finally, in a sarcoma model, recombinant AdV was employed to enable IL-12 expression from DCs with the aim of enhancing cross-priming of tumor-specific CD8+ T cells and tumor rejection [104].

A large body of preclinical studies has made AdV the most frequently investigated OIV in clinical trials (#200), about one-half of which are ongoing or recruiting patients. Of these trials, 13 investigated oAdVs armed with IL-12 and included Ad-hIL12 (constitutive IL-12) (#6), Ad-RTS-hIL12 (tunable IL-12) (#6), and Ad5-yCD/mutTKSR39rep-hIL12 (combinatorial) (#1) against pancreatic, breast, prostatic, and pediatric tumors, glioma, glioblastoma, and melanoma. A completed trial with Ad-IL12 against liver, colorectal, and pancreatic human cancers showed a high safety profile, an increase in tumor infiltration by effector immune cells, yet overall mild antitumor effects [105]. Preliminary results of an ongoing trial with Ad-RTS-hIL12/veledimex (the cytokine inducer) against recurrent high-grade glioma showed the safety and tolerability of the treatment and demonstrated that oAdV elicited a sustained intratumoral immune response. The median overall survival (mOS) was higher in patients treated with the armed virus [90].

4.5   oMeV

An interesting example of the benefits provided by targeted IL-12 delivery through OIVs is offered by the oncolytic measles virus (oMeV). Ungerechts, Engeland, and coworkers engineered a fusion version of murine IL-12 in the virus and named it FmIL-12 MeVac [106]. The backbone was a vaccine strain of MeV (MeVac). In vitro, the infected cells produced large amounts of FmIL-12, up to 2000 ng/mL. In vivo, FmIL-12 MeVac and unarmed MeVac conferred 90 and 40% protection against MC38CEA tumors, respectively. The antitumor efficacy of the IL-12 virus was also superior to that of MeVac expressing anti-PD-L1. The intratumoral administration of FmIL-12 MeVac elicited local and systemic immune responses, documented mainly as intratumoral increases in activated CD8+ T and NK cells, increases in IFNγ and TNFα, splenocyte reactivity to tumor cells, and immune protection from a distant challenge tumor. The unarmed version conferred less protection from a challenge tumor and a very modest or negligible capacity to induce tumor immune-heating. The same authors carried out an interesting comparison of the benefits offered by FmIL-12 MeVac relative to those provided by FmIL-15 MeVac, a virus expressing IL-15 and the sushi-activating portion of its receptor. The former virus was superior in terms of efficacy against primary tumors, even though the two viruses were similar overall with respect to the increase in intratumoral CD8+ T cells and NK cells [107].

4.6   NDV

NDV is an oncolytic virus of bird origin. It replicates in human tumor cells and fails to substantially replicate in noncancerous human cells. An advantage of OIVs of animal origin is the absence of prior immunity in humans, which could neu-tralize the spread of the OIV, particularly upon systemic OIV administration. NDVs also infect dendritic cells [108]. It was initially recognized that an unarmed version of NDV could overcome the immunosuppressive nature of the TME, at least in part, through the induction of IL-12, IFNγ and additional cytokines, and driving a Th1 response [108, 109].

Various groups have independently investigated the benefits of delivering IL-12 intratumorally with the aid of recombinant NDV in murine tumor models [110–112]. In all cases, the IL-12-armed versions were superior to their unarmed NDV counterparts, as assayed in 4 T1 breast cancer, B16 melanoma, and hepatoma models.

Of particular interest is the possibility of encoding both IL-12 and checkpoint inhibitors from the genome of an oncolytic virus to limit the severe adverse effects caused by the systemic administration of immune modulators. An elegant example of this possibility was recently shown with recombinant versions of NVD expressing neutralizing single-chain antibodies (scFvs) against PD-1, PD-L1, or agonistic scFv to the costimulatory CD28 as proteins alone or as fusion proteins with IL-12 (the combination was named checkpoint inhibitor immunocytokines) [112]. The recombinant NDVs were administered to mice bearing B16 melanoma tumors, alone or in combination with systemic anti-CTLA-4. The NDVs expressing the checkpoint inhibitors fused to IL-12 were invariably more potent than their counterparts without IL-12. The IL-12 adjuvant effect converted the highly immunosuppressive and nonresponsive B16 melanoma tumors into immunologically hot tumors, such that the checkpoint inhibitor-immune cytokine synergized with the systemic administration of anti-CTLA-4. Interestingly, these combinations elicited a strong immunotherapeutic effect, highlighted as an abscopal antitumor effect observed on a distant untreated challenge tumor. At present, the strategy of expressing multiple immune-modulatory payloads is being pursued for numerous OIVs [86, 94, 100, 113–116] and by companies, such as Oncorus, Replimune, Turnstone Biologics, Immvira and others, are developing multiply armed OIVs.

4.7 Maraba Virus

Maraba virus (MRB), an oncolytic rhabdovirus of animal origin, selectively infects human tumor cells. MG1 is an IFN-sensitive mutant selected for safety reasons [117]. The major effect of MRBs consists of their ability to elicit antitumor immunity and exert abscopal protection, which makes them among the most effective oncolytic vectors for antitumor vaccination. MRBs are being employed in a prime-boost modality with AdV [118]. Currently, four first-in-human trials are ongoing or recruiting for patients with advanced/metastatic solid tumors, including melanoma, squamous cell skin carcinoma, non-small-cell lung cancer for testing the effect of MRB as a monotherapy or in combination with CPI or adenovirus vaccination.

An IL-12-armed version of MRB MG1 was employed to infect ex vivo autologous tumor cells, which were subsequently administered intraperitoneally as an infected cell vaccine against peritoneal carcinomatosis caused by melanoma B16 or colon carcinoma CT-26 cells in models of metastatic tumors. The treatment promoted the recruitment and activation of NK cells to the peritoneal cavity, causing a reduction in tumor burden and overall improved survival, including complete protection [119].

4.8 Concluding Remarks. *If There Were No IL-12, Someone Would Need to Invent it*

Compelling evidence indicates that IL-12 serves as a potent adjuvant of the immunotherapeutic response elicited by OIVs in murine models of tumors. The adjuvant effect was invariably observed for all the OIVs analyzed. OIVs offered the opportunity to encode IL-12 as a transgene and thus to express the cytokine—be it wt or as recombinant fusion form—at high concentrations in the tumor bed, without most of the adverse side effects and toxicities that have hampered the systemic application of IL-12 in past human trials. In some studies, it was shown that the changes to the TME induced by the IL-12-armed OIVs primed the tumor for the checkpoint blockade therapy, paving the way for the combined IL-12/OIV/checkpoint blockade treatment. Some features of the IL-12-mediated response summon the possibility that IL-12 not only quantitatively boosts the immune response but also increases the antigenic repertoire of B and T cells; this mechanism remains to be analyzed in detail. The highly encouraging results in preclinical models have fueled the translation to the clinic. The extent to which the IL-12/OIV combination holds promise in humans remains to be fully investigated.

Acknowledgments The experimental work carried out in our laboratory reviewed in this chapter was funded by European Research Council (ERC), Advanced Grant number 340060.

References

1. Gately M, Desai B, Wolitzky A, Quinn P, Dwyer C, Podlaski F, Familletti P, Sinigaglia F, Chizonnite R, Gubler U (1991) Regulation of human lymphocyte proliferation by a heterodimeric cytokine, IL-12 (cytotoxic lymphocyte maturation factor). J Immunol 147(3):874–882
2. Kobayashi M, Fitz L, Ryan M, Hewick RM, Clark SC, Chan S, Loudon R, Sherman F, Perussia B, Trinchieri G (1989) Identification and purification of natural killer cell stimulatory factor (NKSF), a cytokine with multiple biologic effects on human lymphocytes. J Exp Med 170(3):827–845
3. Stern AS, Podlaski FJ, Hulmes JD, Pan Y, Quinn PM, Wolitzky A, Familletti PC, Stremlo DL, Truitt T, Chizzonite R (1990) Purification to homogeneity and partial characterization of cytotoxic lymphocyte maturation factor from human B-lymphoblastoid cells. Proc Natl Acad Sci 87(17):6808–6812
4. Vignali DA, Kuchroo VK (2012) IL-12 family cytokines: immunological playmakers. Nat Immunol 13(8):722
5. Mal X, Trinchieri G (2001) Regulation of interleukin-12 production in antigen-presenting cells. Adv Immunol Actions 79:55–92
6. Airoldi I, Gri G, Marshall JD, Corcione A, Facchetti P, Guglielmino R, Trinchieri G, Pistoia V (2000) Expression and function of IL-12 and IL-18 receptors on human tonsillar B cells. J Immunol 165(12):6880–6888
7. Grohmann U, Belladonna ML, Bianchi R, Orabona C, Ayroldi E, Fioretti MC, Puccetti P (1998) IL-12 acts directly on DC to promote nuclear localiza-

tion of NF-κB and primes DC for IL-12 production. Immunity 9(3):315–323

8. Grohmann U, Belladonna ML, Vacca C, Bianchi R, Fallarino F, Orabona C, Fioretti MC, Puccetti P (2001) Positive regulatory role of IL-12 in macrophages and modulation by IFN-γ. J Immunol 167(1):221–227

9. Presky DH, Yang H, Minetti LJ, Chua AO, Nabavi N, Wu C-Y, Gately MK, Gubler U (1996) A functional interleukin 12 receptor complex is composed of two β-type cytokine receptor subunits. Proc Natl Acad Sci 93(24):14002–14007

10. Afonso L, Scharton TM, Vieira LQ, Wysocka M, Trinchieri G, Scott P (1994) The adjuvant effect of interleukin-12 in a vaccine against Leishmania major. Science 263(5144):235–237

11. Hall SS (1994) IL-12 holds promise against cancer, glimmer of AIDS hope. Science 263(5154):1685–1687

12. Heaton KM, Grimm EA (1993) Cytokine combinations in immunotherapy for solid tumors: a review. Cancer Immunol Immunother 37(4):213–219

13. Trinchieri G (2003) Interleukin-12 and the regulation of innate resistance and adaptive immunity. Nat Rev Immunol 3(2):133

14. Smits HH, van Rietschoten JG, Hilkens CM, Sayilir R, Stiekema F, Kapsenberg ML, Wierenga EA (2001) IL-12-induced reversal of human Th2 cells is accompanied by full restoration of IL-12 responsiveness and loss of GATA-3 expression. Eur J Immunol 31(4):1055–1065

15. Perussia B, Chan SH, D'Andrea A, Tsuji K, Santoli D, Pospisil M, Young D, Wolf SF, Trinchieri G (1992) Natural killer (NK) cell stimulatory factor or IL-12 has differential effects on the proliferation of TCR-alpha beta+, TCR-gamma delta+ T lymphocytes, and NK cells. J Immunol 149(11):3495–3502

16. Watkins SK, Egilmez NK, Suttles J, Stout RD (2007) IL-12 rapidly alters the functional profile of tumor-associated and tumor-infiltrating macrophages in vitro and in vivo. J Immunol 178(3):1357–1362

17. Fukao T, Frucht DM, Yap G, Gadina M, O'Shea JJ, Koyasu S (2001) Inducible expression of Stat4 in dendritic cells and macrophages and its critical role in innate and adaptive immune responses. J Immunol 166(7):4446–4455

18. Birbrair A, Zhang T, Wang Z-M, Messi ML, Olson JD, Mintz A, Delbono O (2014) Type-2 pericytes participate in normal and tumoral angiogenesis. Am J Physiol Cell Physiol 307(1):C25–C38

19. Airoldi I, Di Carlo E, Cocco C, Caci E, Cilli M, Sorrentino C, Sozzi G, Ferrini S, Rosini S, Bertolini G, Truini M, Grossi F, Galietta LJV, Ribatti D, Pistoia V (2009) IL-12 can target human lung adenocarcinoma cells and normal bronchial epithelial cells surrounding tumor lesions. PLos One 4(7):e6119

20. Strasly M, Cavallo F, Geuna M, Mitola S, Colombo MP, Forni G, Bussolino F (2001) IL-12 inhibition of endothelial cell functions and angiogenesis

depends on lymphocyte-endothelial cell cross-talk. J Immunol 166(6):3890–3899

21. Yago T, Tsukuda M, Fukushima H, Yamaoka H, Kurata-Miura K, Nishi T, Minami M (1998) IL-12 promotes the adhesion of NK cells to endothelial selectins under flow conditions. J Immunol 161(3):1140–1145

22. Colombo MP, Trinchieri G (2002) Interleukin-12 in anti-tumor immunity and immunotherapy. Cytokine Growth Factor Rev 13(2):155–168

23. Cohen J (1995) IL-12 deaths: explanation and a puzzle. Science 270(5238):908–908

24. Del Vecchio M, Bajetta E, Canova S, Lotze MT, Wesa A, Parmiani G, Anichini A (2007) Interleukin-12: biological properties and clinical application. Clin Cancer Res 13(16):4677–4685

25. Leonard JP, Sherman ML, Fisher GL, Buchanan LJ, Larsen G, Atkins MB, Sosman JA, Dutcher JP, Vogelzang NJ, Ryan JL (1997) Effects of single-dose interleukin-12 exposure on interleukin-12–associated toxicity and interferon-γ production. Blood 90(7):2541–2548

26. Motzer RJ, Rakhit A, Schwartz LH, Olencki T, Malone TM, Sandstrom K, Nadeau R, Parmar H, Bukowski R (1998) Phase I trial of subcutaneous recombinant human interleukin-12 in patients with advanced renal cell carcinoma. Clin Cancer Res 4(5):1183–1191

27. Caruso M, Pham-Nguyen K, Kwong Y-L, Xu B, Kosai K-I, Finegold M, Woo S, Chen S-H (1996) Adenovirus-mediated interleukin-12 gene therapy for metastatic colon carcinoma. Proc Natl Acad Sci 93(21):11302–11306

28. Qian C, Liu XY, Prieto J (2006) Therapy of cancer by cytokines mediated by gene therapy approach. Cell Res 16(2):182

29. Barrett JA, Cai H, Miao J, Khare PD, Gonzalez P, Dalsing-Hernandez J, Sharma G, Chan T, Cooper LJ, Lebel F (2018) Regulated intratumoral expression of IL-12 using a RheoSwitch therapeutic system®(RTS®) gene switch as gene therapy for the treatment of glioma. Cancer Gene Ther 25(5):106

30. Rudman SM, Jameson MB, McKeage MJ, Savage P, Jodrell DI, Harries M, Acton G, Erlandsson F, Spicer JF (2011) A phase 1 study of AS1409, a novel antibody-cytokine fusion protein, in patients with malignant melanoma or renal cell carcinoma. Clin Cancer Res 17(7):1998–2005

31. Strauss J, Heery CR, Kim JW, Jochems C, Donahue RN, Montgomery AS, McMahon S, Lamping E, Marté JL, Madan RA (2019) First-in-human phase I trial of a tumor-targeted cytokine (NHS-IL12) in subjects with metastatic solid tumors. Clin Cancer Res 25(1):99–109

32. Kueberuwa G, Kalaitsidou M, Cheadle E, Hawkins RE, Gilham DE (2018) CD19 CAR T cells expressing IL-12 eradicate lymphoma in fully lymphoreplete mice through induction of host immunity. Mol Ther Oncolytics 8:41–51

33. Zhang L, Morgan RA, Beane JD, Zheng Z, Dudley ME, Kassim SH, Nahvi AV, Ngo LT, Sherry RM, Phan GQ (2015) Tumor-infiltrating lymphocytes genetically engineered with an inducible gene encoding interleukin-12 for the immunotherapy of metastatic melanoma. Clin Cancer Res 21(10):2278–2288

34. Martuza RL, Malick A, Markert JM, Ruffner KL, Coen DM (1991) Experimental therapy of human glioma by means of a genetically engineered virus mutant. Science 252(5007):854–856

35. Vile RG (2018) The immune system in oncolytic immunovirotherapy: gospel, schism and heresy. Mol Ther 26(4):942–946

36. Andtbacka RH (2016) The role of talimogene laherparepvec (T-VEC) in the age of checkpoint inhibitors. Clin Adv Hematol Oncol 14(8):576

37. Coffin RS (2015) From virotherapy to oncolytic immunotherapy: where are we now? Curr Opin Virol 13:93–100

38. Russell SJ, Barber GN (2018) Oncolytic viruses as antigen-agnostic cancer vaccines. Cancer Cell 33(4):599–605

39. Kaufman HL, Kim DW, DeRaffele G, Mitcham J, Coffin RS, Kim-Schulze S (2010) Local and distant immunity induced by intralesional vaccination with an oncolytic herpes virus encoding GM-CSF in patients with stage IIIc and IV melanoma. Ann Surg Oncol 17(3):718–730

40. Zamarin D, Holmgaard RB, Subudhi SK, Park JS, Mansour M, Palese P, Merghoub T, Wolchok JD, Allison JP (2014) Localized oncolytic virotherapy overcomes systemic tumor resistance to immune checkpoint blockade immunotherapy. Sci Trans Med 6(226):226ra232

41. Biotech J (2019) Checkpoint inhibitors go viral. Nat Biotechnol 37(1):13

42. Lawler SE, Speranza M-C, Cho C-F, Chiocca EA (2017) Oncolytic viruses in cancer treatment: a review. JAMA Oncol 3(6):841–849

43. Ribas A, Dummer R, Puzanov I, VanderWalde A, Andtbacka RH, Michielin O, Olszanski AJ, Malvehy J, Cebon J, Fernandez E (2017) Oncolytic virotherapy promotes intratumoral T cell infiltration and improves anti-PD-1 immunotherapy. Cell 170(6):1109–1119.e1110

44. Gujar S, Pol JG, Kim Y, Lee PW, Kroemer G (2018) Antitumor benefits of antiviral immunity: an underappreciated aspect of oncolytic virotherapies. Trends Immunol 39(3):209–221

45. Pearl TM, Markert JM, Cassady KA, Ghonime MG (2019) Oncolytic virus-based cytokine expression to improve immune activity in brain and solid tumors. Mol Ther Oncolytics 13:14

46. Toda M, Martuza RL, Kojima H, Rabkin SD (1998) In situ cancer vaccination: an IL-12 defective vector/replication-competent herpes simplex virus combination induces local and systemic antitumor activity. J Immunol 160(9):4457–4464

47. Cheema TA, Wakimoto H, Fecci PE, Ning J, Kuroda T, Jeyaretna DS, Martuza RL, Rabkin SD (2013) Multifaceted oncolytic virus therapy for glioblastoma in an immunocompetent cancer stem cell model. Proc Natl Acad Sci 110(29):12006–12011

48. Bennett JJ, Malhotra S, Wong RJ, Delman K, Zager J, St-Louis M, Johnson P, Fong Y (2001) Interleukin 12 secretion enhances antitumor efficacy of oncolytic herpes simplex viral therapy for colorectal cancer. Ann Surg 233(6):819

49. Jarnagin W, Zager J, Klimstra D, Delman K, Malhotra S, Ebright M, Little S, DeRubertis B, Stanziale S, Hezel M (2003) Neoadjuvant treatment of hepatic malignancy: an oncolytic herpes simplex virus expressing IL-12 effectively treats the parent tumor and protects against recurrence-after resection. Cancer Gene Ther 10(3):215

50. Wong RJ, Patel SG, Kim S-H, DeMatteo RP, Malhotra S, Bennett JJ, St-Louis M, Shah JP, Johnson PA, Fong Y (2001) Cytokine gene transfer enhances herpes oncolytic therapy in murine squamous cell carcinoma. Hum Gene Ther 12(3):253–265

51. Varghese S, Rabkin S, Liu R, Nielsen P, Ipe T, Martuza R (2006b) Enhanced therapeutic efficacy of IL-12, but not GM-CSF, expressing oncolytic herpes simplex virus for transgenic mouse derived prostate cancers. Cancer Gene Ther 13(3):253

52. Varghese S, Rabkin SD, Nielsen PG, Wang W, Martuza RL (2006a) Systemic oncolytic herpes virus therapy of poorly immunogenic prostate cancer metastatic to lung. Clin Cancer Res 12(9):2919–2927

53. Wong RJ, Chan M-K, Yu Z, Ghossein RA, Ngai I, Adusumilli PS, Stiles BM, Shah JP, Singh B, Fong Y (2004a) Angiogenesis inhibition by an oncolytic herpes virus expressing interleukin 12. Clin Cancer Res 10(13):4509–4516

54. Wong RJ, Chan M-K, Yu Z, Kim TH, Bhargava A, Stiles BM, Horsburgh BC, Shah JP, Ghossein RA, Singh B (2004b) Effective intravenous therapy of murine pulmonary metastases with an oncolytic herpes virus expressing interleukin 12. Clin Cancer Res 10(1):251–259

55. Derubertis B, Stiles B, Bhargava A, Gusani N, Hezel M, D'Angelica M, Fong Y (2007) Cytokine-secreting herpes viral mutants effectively treat tumor in a murine metastatic colorectal liver model by oncolytic and T-cell-dependent mechanisms. Cancer Gene Ther 14(6):590

56. Varghese S, Rabkin SD, Nielsen GP, MacGarvey U, Liu R, Martuza RL (2007) Systemic therapy of spontaneous prostate cancer in transgenic mice with oncolytic herpes simplex viruses. Cancer Res 67(19):9371–9379

57. Parker JN, Gillespie GY, Love CE, Randall S, Whitley RJ, Markert JM (2000) Engineered herpes simplex virus expressing IL-12 in the treatment of experimental murine brain tumors. Proc Natl Acad Sci 97(5):2208–2213

58. Cody JJ, Scaturro P, Cantor AB, Yancey Gillespie G, Parker JN, Markert JM (2012) Preclinical evaluation of oncolytic Δγ134. 5 herpes simplex virus express-

ing interleukin-12 for therapy of breast cancer brain metastases. Int J Breast Cancer 2012:628697

59. Hellums EK, Markert JM, Parker JN, He B, Perbal B, Roizman B, Whitley RJ, Langford CP, Bharara S, Gillespie GY (2005) Increased efficacy of an interleukin-12-secreting herpes simplex virus in a syngeneic intracranial murine glioma model. Neuro-Oncology 7(3):213–224

60. Bauer DF, Pereboeva L, Gillespie GY, Cloud GA, Elzafarany O, Langford C, Markert JM (2016) Effect of HSV-IL12 loaded tumor cell-based vaccination in a mouse model of high-grade neuroblastoma. J Immunol Res 2016:2568125

61. Waters AM, Stafman LL, Garner EF, Mruthyunjayappa S, Stewart JE, Friedman GK, Coleman JM, Markert JM, Gillespie GY, Beierle EA (2016) Effect of repeat dosing of engineered oncolytic herpes simplex virus on preclinical models of rhabdomyosarcoma. Transl Oncol 9(5):419–430

62. Ring EK, Li R, Moore BP, Nan L, Kelly VM, Han X, Beierle EA, Markert JM, Leavenworth JW, Gillespie GY (2017) Newly characterized murine undifferentiated sarcoma models sensitive to virotherapy with oncolytic HSV-1 M002. Mol Ther Oncolytics 7:27–36

63. Friedman GK, Bernstock JD, Chen D, Nan L, Moore BP, Kelly VM, Youngblood SL, Langford CP, Han X, Ring EK (2018) Enhanced sensitivity of patient-derived pediatric high-grade brain tumor xenografts to oncolytic hsv-1 virotherapy correlates with nectin-1 expression. Sci Rep 8(1):13930

64. Friedman GK, Moore BP, Nan L, Kelly VM, Etminan T, Langford CP, Xu H, Han X, Markert JM, Beierle EA (2015) Pediatric medulloblastoma xenografts including molecular subgroup 3 and CD133+ and CD15+ cells are sensitive to killing by oncolytic herpes simplex viruses. Neuro-Oncology 18(2):227–235

65. Markert JM, Cody JJ, Parker JN, Coleman JM, Price KH, Kern ER, Quenelle DC, Lakeman AD, Schoeb TR, Palmer CA (2012) Preclinical evaluation of a genetically engineered herpes simplex virus expressing interleukin-12. J Virol 86(9):5304–5313

66. Roth JC, Cassady KA, Cody JJ, Parker JN, Price KH, Coleman JM, Peggins JO, Noker PE, Powers NW, Grimes SD (2014) Evaluation of the safety and biodistribution of M032, an attenuated herpes simplex virus type 1 expressing hIL-12, after intracerebral administration to aotus nonhuman primates. Hum Gene Ther Clin Dev 25(1):16–27

67. Patel DM, Foreman PM, Nabors LB, Riley KO, Gillespie GY, Markert JM (2016) Design of a phase I clinical trial to evaluate M032, a genetically engineered HSV-1 expressing IL-12, in patients with recurrent/progressive glioblastoma multiforme, anaplastic astrocytoma, or gliosarcoma. Hum Gene Ther Clin Dev 27(2):69–78

68. Menotti L, Cerretani A, Hengel H, Campadelli-Fiume G (2008) Construction of a fully retargeted herpes simplex virus 1 recombinant capable of enter-ing cells solely via human epidermal growth factor receptor 2. J Virol 82(20):10153–10161

69. Uchida H, Marzulli M, Nakano K, Goins WF, Chan J, Hong C-S, Mazzacurati L, Yoo JY, Haseley A, Nakashima HJMT (2013) Effective treatment of an orthotopic xenograft model of human glioblastoma using an EGFR-retargeted oncolytic herpes simplex virus. Mol Ther 21(3):561–569

70. Zhou G, Roizman B (2007) Separation of receptor-binding and profusogenic domains of glycoprotein D of herpes simplex virus 1 into distinct interacting proteins. Proc Natl Acad Sci U S A 104(10):4142–4146

71. Slamon DJ, Clark GM, Wong SG, Levin WJ, Ullrich A, McGuire WL (1987) Human breast cancer: correlation of relapse and survival with amplification of the HER-2/neu oncogene. Science 235(4785):177–182

72. Yan M, Schwaederle M, Arguello D, Millis SZ, Gatalica Z, Kurzrock R (2015) HER2 expression status in diverse cancers: review of results from 37,992 patients. Cancer Metastasis Rev 34(1):157–164

73. Leoni V, Vannini A, Gatta V, Rambaldi J, Sanapo M, Barboni C, Zaghini A, Nanni P, Lollini P-L, Casiraghi C (2018) A fully-virulent retargeted oncolytic HSV armed with IL-12 elicits local immunity and vaccine therapy towards distant tumors. PLoS Pathog 14(8):e1007209

74. Alessandrini F, Menotti L, Avitabile E, Appolloni I, Ceresa D, Marubbi D, Campadelli-Fiume G, Malatesta P (2019) Eradication of glioblastoma by immuno-virotherapy with a retargeted oncolytic HSV in a preclinical model. Oncogene 38(23):4467

75. Wold WSM, Toth K (2013) Adenovirus vectors for gene therapy, vaccination and cancer gene therapy. Curr Gene Ther 13(6):421–433

76. Wu Q, Moyana T, Xiang J (2001) Cancer gene therapy by adenovirus-mediated gene transfer. Curr Gene Ther 1(1):101–122

77. Bramson J, Hitt M, Addison C, Muller W, Gauldie J, Graham F (1996) Direct intratumoral injection of an adenovirus expressing interleukin-12 induces regression and long-lasting immunity that is associated with highly localized expression of interleukin-12. Hum Gene Ther 7(16):1995–2002

78. Barajas M, Mazzolini G, Genové G, Bilbao R, Narvaiza I, Schmitz V, Sangro B, Melero I, Qian C, Prieto J (2001) Gene therapy of orthotopic hepatocellular carcinoma in rats using adenovirus coding for interleukin 12. Hepatology 33(1):52–61

79. Mazzolini G, Qian C, Xie X, Sun Y, Lasarte JJ, Drozdzik M, Prieto J (1999) Regression of colon cancer and induction of antitumor immunity by intratumoral injection of adenovirus expressing interleukin-12. Cancer Gene Ther 6(6):514

80. Liu Y, Ehtesham M, Samoto K, Wheeler CJ, Thompson RC, Villarreal LP, Black KL, John SY (2002) In situ adenoviral interleukin 12 gene transfer confers potent and long-lasting cytotoxic immunity in glioma. Cancer Gene Ther 9(1):9

81. Nasu Y, Bangma C, Hull G, Lee H, Hu J, Wang J, McCurdy M, Shimura S, Yang G, Timme T (1999)

Adenovirus-mediated interleukin-12 gene therapy for prostate cancer: suppression of orthotopic tumor growth and pre-established lung metastases in an orthotopic model. Gene Ther 6(3):338–349

82. Zhang R, DeGroot LJ (2000) Genetic immunotherapy of established tumours with adenoviral vectors transducing murine interleukin-12 (mIL-12) stimulate in a rat medullary thyroid carcinoma model. Clin Endocrinol 52(6):687–694

83. Zhang R, DeGroot LJ (2003) Gene therapy of a rat follicular thyroid carcinoma model with adenoviral vectors transducing murine interleukin-12. Endocrinology 144(4):1393–1398

84. Bortolanza S, Bunuales M, Otano I, Gonzalez-Aseguinolaza G, Ortiz-de-Solorzano C, Perez D, Prieto J, Hernandez-Alcoceba R (2009) Treatment of pancreatic cancer with an oncolytic adenovirus expressing interleukin-12 in Syrian hamsters. Mol Ther 17(4):614–622

85. Endo Y, Sakai R, Ouchi M, Onimatsu H, Hioki M, Kagawa S, Uno F, Watanabe Y, Urata Y, Tanaka N (2008) Virus-mediated oncolysis induces danger signal and stimulates cytotoxic T-lymphocyte activity via proteasome activator upregulation. Oncogene 27(17):2375

86. Huang J-H, Zhang S-N, Choi K-J, Choi I-K, Kim J-H, Lee M, Kim H, Yun C-O (2010) Therapeutic and tumor-specific immunity induced by combination of dendritic cells and oncolytic adenovirus expressing IL-12 and 4-1BBL. Mol Ther 18(2):264–274

87. Freytag S, Barton K, Zhang Y (2013) Efficacy of oncolytic adenovirus expressing suicide genes and interleukin-12 in preclinical model of prostate cancer. Gene Ther 20(12):1131

88. Choi K-J, Zhang S-N, Choi I-K, Kim J-S, Yun C-O (2012) Strengthening of antitumor immune memory and prevention of thymic atrophy mediated by adenovirus expressing IL-12 and GM-CSF. Gene Ther 19(7):711

89. Divino CM, Chen S-H, Yang W, Thung S, Brower ST, Woo S (2000) Anti-tumor immunity induced by interleukin-12 gene therapy in a metastatic model of breast cancer is mediated by natural killer cells. Breast Cancer Res Treat 60(2):129–134

90. Chiocca EA, John SY, Lukas RV, Solomon IH, Ligon KL, Nakashima H, Triggs DA, Reardon DA, Wen P, Stopa BM (2019) Regulatable interleukin-12 gene therapy in patients with recurrent high-grade glioma: results of a phase 1 trial. Sci Trans Med 11(505):eaaw5680

91. Komita H, Zhao X, Katakam AK, Kumar P, Kawabe M, Okada H, Braughler JM, Storkus WJ (2009) Conditional interleukin-12 gene therapy promotes safe and effective antitumor immunity. Cancer Gene Ther 16(12):883

92. Wang P, Li X, Wang J, Gao D, Li Y, Li H, Chu Y, Zhang Z, Liu H, Jiang G (2017) Re-designing Interleukin-12 to enhance its safety and potential as an anti-tumor immunotherapeutic agent. Nat Commun 8(1):1395

93. Poutou J, Bunuales M, Gonzalez-Aparicio M, Garcia-Aragoncillo E, Quetglas JI, Casado R, Bravo-Perez C, Alzuguren P, Hernandez-Alcoceba R (2015) Safety and antitumor effect of oncolytic and helper-dependent adenoviruses expressing interleukin-12 variants in a hamster pancreatic cancer model. Gene Ther 22(9):696

94. Choi I-K, Li Y, Oh E, Kim J, Yun C-O (2013) Oncolytic adenovirus expressing IL-23 and p35 elicits IFN-γ-and TNF-α-co-producing T cell-mediated antitumor immunity. PLoS One 8(7):e67512

95. Choi I, Lee J, Zhang S, Park J, Lee K-M, Sonn C, Yun CO (2011) Oncolytic adenovirus co-expressing IL-12 and IL-18 improves tumor-specific immunity via differentiation of T cells expressing IL-12Rβ 2 or IL-18Rα. Gene Ther 18(9):898–909

96. Kim W, Seong J, Oh HJ, Koom WS, Choi K-J, Yun C-O (2011) A novel combination treatment of armed oncolytic adenovirus expressing IL-12 and GM-CSF with radiotherapy in murine hepatocarcinoma. J Radiat Res 52(5):646–654

97. Lee Y-S, Kim J-H, Choi K-J, Choi I-K, Kim H, Cho S, Cho BC, Yun C-O (2006) Enhanced antitumor effect of oncolytic adenovirus expressing interleukin-12 and B7-1 in an immunocompetent murine model. Clin Cancer Res 12(19):5859–5868

98. Pützer BM, Hitt M, Muller WJ, Emtage P, Gauldie J, Graham FL (1997) Interleukin 12 and B7-1 costimulatory molecule expressed by an adenovirus vector act synergistically to facilitate tumor regression. Proc Natl Acad Sci 94(20):10889–10894

99. Zhang S-N, Choi I-K, Huang J-H, Yoo J-Y, Choi K-J, Yun C-O (2011) Optimizing DC vaccination by combination with oncolytic adenovirus coexpressing IL-12 and GM-CSF. Mol Ther 19(8):1558–1568

100. Ahn HM, Hong J, Yun C-O (2016) Oncolytic adenovirus coexpressing interleukin-12 and shVEGF restores antitumor immune function and enhances antitumor efficacy. Oncotarget 7(51):84965

101. Oh E, Choi I-K, Hong J, Yun C-O (2017) Oncolytic adenovirus coexpressing interleukin-12 and decorin overcomes Treg-mediated immunosuppression inducing potent antitumor effects in a weakly immunogenic tumor model. Oncotarget 8(3):4730

102. Shaw AR, Porter C, Watanabe N, Tanoue K, Sikora A, Gottschalk S, Brenner M, Suzuki M (2017a) Adenovirotherapy delivering cytokine and checkpoint inhibitor augments chimeric antigen receptor T-cell against metastatic head and neck cancer. Mol Ther 25(11):2440–2451

103. Shaw AR, Porter CE, Watanabe N, Tanoue K, Sikora A, Gottschalk S, Brenner MK, Suzuki M (2017b) Adenovirotherapy delivering cytokine and checkpoint inhibitor augments CAR T cells against metastatic head and neck cancer. Mol Ther 25(11):2440–2451

104. Tatsumi T, Huang J, Gooding WE, Gambotto A, Robbins PD, Vujanovic NL, Alber SM, Watkins SC, Okada H, Storkus WJ (2003) Intratumoral delivery of dendritic cells engineered to secrete both inter-

leukin (IL)-12 and IL-18 effectively treats local and distant disease in association with broadly reactive Tc1-type immunity. Cancer Res 63(19):6378–6386

105. Sangro B, Mazzolini G, Ruiz J, Herraiz M, Quiroga J, Herrero I, Benito A, Larrache J, Pueyo J, Subtil JC (2004) Phase I trial of intratumoral injection of an adenovirus encoding interleukin-12 for advanced digestive tumors. J Clin Oncol 22(8):1389–1397

106. Veinalde R, Grossardt C, Hartmann L, Bourgeois-Daigneault M-C, Bell JC, Jäger D, von Kalle C, Ungerechts G, Engeland CE (2017) Oncolytic measles virus encoding interleukin-12 mediates potent antitumor effects through T cell activation. Onco Targets Ther 6(4):e1285992

107. Backhaus PS, Veinalde R, Hartmann L, Dunder JE, Jeworowski LM, Albert J, Hoyler B, Poth T, Jäger D, Ungerechts G (2019) Immunological effects and viral gene expression determine the efficacy of oncolytic measles vaccines encoding IL-12 or IL-15 agonists. Viruses 11(10):914

108. Tan L, Zhang Y, Qiao C, Yuan Y, Sun Y, Qiu X, Meng C, Song C, Liao Y, Munir M (2018) NDV entry into dendritic cells through macropinocytosis and suppression of T lymphocyte proliferation. Virology 518:126–135

109. Lam HY, Yusoff K, Yeap SK, Subramani T, Abd-Aziz S, Omar AR, Alitheen NB (2014) Immunomodulatory effects of Newcastle disease virus AF2240 strain on human peripheral blood mononuclear cells. Int J Med Sci 11(12):1240

110. Amin ZM, Ani MAC, Tan SW, Yeap SK, Alitheen NB, Najmuddin SUFS, Kalyanasundram J, Chan SC, Veerakumarasivam A, Chia SL (2019) Evaluation of a recombinant newcastle disease virus expressing human IL12 against human breast cancer. Sci Rep 9(1):1–10

111. Ren G, Tian G, Liu Y, He J, Gao X, Yu Y, Liu X, Zhang X, Sun T, Liu S (2016) Recombinant newcastle disease virus encoding IL-12 and/or IL-2 as potential candidate for hepatoma carcinoma therapy. Technol Cancer Res Treat 15(5):NP83–NP94

112. Vijayakumar G, McCroskery S, Palese P (2020) Engineering newcastle disease virus as oncolytic vector for intratumoral delivery of immune checkpoint inhibitors and immunocytokines. J Virol 94(3):e01677–e01619

113. Fukuhara H, Ino Y, Kuroda T, Martuza RL, Todo T (2005) Triple gene-deleted oncolytic herpes simplex virus vector double-armed with interleukin 18 and soluble B7-1 constructed by bacterial artificial chromosome–mediated system. Cancer Res 65(23):10663–10668

114. Parker JN, Meleth S, Hughes KB, Gillespie GY, Whitley RJ, Markert JM (2005) Enhanced inhibition of syngeneic murine tumors by combinatorial therapy with genetically engineered HSV-1 expressing CCL2 and IL-12. Cancer Gene Ther 12(4):359

115. Thomas S, Kuncheria L, Roulstone V, Kyula JN, Mansfield D, Bommareddy PK, Smith H, Kaufman HL, Harrington KJ, Coffin RS (2019) Development of a new fusion-enhanced oncolytic immunotherapy platform based on herpes simplex virus type 1. J Immunother Cancer 7(1):1–17

116. Yan R, Zhou X, Chen X, Liu X, Tang Y, Ma J, Wang L, Liu Z, Zhan B, Chen H, Wang J, Zou W, Xu H, Lu R, Ni1 D, Roizman B, Zhou GG (2019) Enhancement of Oncolytic Activity of oHSV Expressing IL-12 and Anti PD-1 Antibody by Concurrent Administration of Exosomes Carrying CTLA-4 miRNA. Immunotherapy 5(1). https://www.longdom.org/open-access/enhancement-of-oncolytic-activity-of-ohsv-expressing-il12-and-anti-pd1-antibody-by-concurrent-administration-ofexosomes-carrying--44157.html

117. Brun J, McManus D, Lefebvre C, Hu K, Falls T, Atkins H, Bell JC, McCart JA, Mahoney D, Stojdl DF (2010) Identification of genetically modified Maraba virus as an oncolytic rhabdovirus. Mol Ther 18(8):1440–1449

118. Pol JG, Zhang L, Bridle BW, Stephenson KB, Rességuier J, Hanson S, Chen L, Kazdhan N, Bramson JL, Stojdl DF (2014) Maraba virus as a potent oncolytic vaccine vector. Mol Ther 22(2):420–429

119. Alkayyal AA, Tai L-H, Kennedy MA, de Souza CT, Zhang J, Lefebvre C, Sahi S, Ananth AA, Mahmoud AB, Makrigiannis AP (2017) NK-cell recruitment is necessary for eradication of peritoneal carcinomatosis with an IL12-expressing Maraba virus cellular vaccine. Cancer Immunol Res 5(3):211–221

IL-22 Signaling in the Tumor Microenvironment

5

Runqiu Jiang and Beicheng Sun

Abstract

Interleukin (IL)-22 belongs to the IL-10 cytokine family which performs biological functions by binding to heterodimer receptors comprising a type 1 receptor chain (R1) and a type 2 receptor chain (R2). IL-22 is mainly derived from CD4+ helper T cells, CD8+ cytotoxic T cells, innate lymphocytes, and natural killer T cells. It can activate downstream signaling pathways such as signal transducer and activator of transcription (STAT)1/3/5, nuclear factor kappa-light-chain-enhancer of activated B cells (NF-κB), mitogen-activated protein kinase (MAPK), and phosphoinositide 3-kinase (PI3K)-protein kinase B (AKT)-mammalian target of rapamycin (mTOR) through these heterodimer receptors. Although IL-22 is produced by immune cells, its specific receptor IL-22R1 is selectively expressed in nonimmune cells, such as hepatocytes, colonic epithelial cells, and pancreatic epithelial cells (Jiang et al. Hepatology 54(3):900–9, 2011; Jiang et al. BMC Cancer 13:59, 2013; Curd et al. Clin Exp Immunol 168(2):192–9, 2012). Immune cells do not respond to IL-22 stimulation directly within tumors, reports from different groups have revealed that IL-22 can indirectly regulate the tumor microenvironment (TME). In the present chapter, we discuss the roles of IL-22 in malignant cells and immunocytes within the TME, meanwhile, the potential roles of IL-22 as a target for drug discovery will be discussed.

Keywords

IL-22 · Tumor microenvironment · Liver cancer · Colorectal cancer · Gastric cancer · Pancreatic cancer · Lung cancer · Breast cancer · Leukemia · Lymphoma · Drug discovery · T cell · Innate lymphocyte · NKT cell · JAK/STAT

R. Jiang
Department of Hepatobiliary Surgery, The Affiliated Drum Tower Hospital of Nanjing University Medical School, Nanjing, People's Republic of China

Medical School of Nanjing University, Nanjing, People's Republic of China

B. Sun (✉)
Department of Hepatobiliary Surgery, The Affiliated Drum Tower Hospital of Nanjing University Medical School, Nanjing, People's Republic of China
e-mail: sunbc@nju.edu.cn

Interleukin (IL)-22 belongs to the IL-10 cytokine family, including IL-10, IL-19, IL-20, IL-22, IL-24, IL-26, IL-28 (α and β), and IL-29 [1]. These cytokines share similarities in their gene

A. Birbrair (ed.), *Tumor Microenvironment*, Advances in Experimental Medicine and Biology 1290, https://doi.org/10.1007/978-3-030-55617-4_5

sequences, protein structures, and receptors. Cytokines belonging to the IL-10 family perform biological functions by binding to heterodimer receptors, which are transmembrane receptor complexes comprising a type 1 receptor chain (R1) and a type 2 receptor chain (R2). Since there are only four types of R1 receptors and two types of R2 receptors, cytokines in the IL-10 family share receptor chains [2]. For instance, the IL-10R2 receptor chain is a shared receptor for IL-10, IL-22, IL-26, IL-28, and IL-29. More interestingly, not only can these cytokines share a single receptor, some cytokines in the IL-10 family even share the same receptor complex. For example, the receptor complex composed of IL-20R1 and IL-20R2 simultaneously mediates signal transduction of IL-19, IL-20, and IL-24 [3]. Although the IL-10 family cytokines share similar receptors, their biological roles are distinct [4, 5].

Endogenous IL-22 is mainly derived from CD4+ helper T cells, CD8+ cytotoxic T cells, innate lymphocytes, and natural killer T cells [6, 7]. It can activate downstream signaling pathways such as signal transducer and activator of transcription (STAT)1/3/5, nuclear factor kappa-light-chain-enhancer of activated B cells (NF-κB), mitogen-activated protein kinase (MAPK), and phosphoinositide 3-kinase (PI3K)-protein kinase B (AKT)-mammalian target of rapamycin (mTOR) through these heterodimer receptors, including IL-22R1 and IL-10R2 [8, 9]. Among these signaling pathways, IL-22/STAT3 is particularly pronounced because of its various biological functions, including inflammation, mitosis, and promotion of cell proliferation, but inhibition of apoptosis [10]. Thus, IL-22 has been defined as a tumor-promoting cytokine. Although IL-22 is produced by immune cells, its specific receptor IL-22R1 is selectively expressed in non-immune cells, such as hepatocytes, colonic epithelial cells, and pancreatic epithelial cells [11–13]. Immune cells do not respond to IL-22 stimulation directly within tumors; reports from different groups have revealed that IL-22 can indirectly regulate the tumor microenvironment (TME). In the present chapter, we discuss the role of IL-22 in malignant cells and immunocytes within the TME.

5.1 IL-22 and Immunocytes Within the TME

Although the IL-22 receptor is strictly expressed on the surface of epithelial cells and cannot directly mediate the immune system, it is inevitable that IL-22 affects the immune system as an important immune regulator, mainly via two manners. First, by stimulating epithelial or tumor cells to produce cytokines that directly modulate immune cells, IL-22 promotes immune escape by inducing production of immunosuppressive cytokines. For example, IL-22 can enhance the production and secretion of IL-6 and IL-8, which are two important inflammatory factors in the TME [14]. IL-22 enhances transforming growth factor beta (TGF-β) (TGF-) and IL-10 production from pancreatic cancer cells to maintain immunosuppression by further inhibiting interferon-gamma (IFN-γ) production in T cells [13]. Second, IL-22 regulates adaptive immunity by binding to its receptors on lymphoepithelial cells. IL-22 can promote the proliferation and survival of lymphoepithelial cells, thus playing an important role in thymic injury regeneration and T-cell development [15]. However, there are few reports on the direct effect of IL-22 on the immune cells in the TME, which may be a direction of future research. More studies have focused on the effects of IL-22 on tumor cells (Fig. 5.1).

5.2 IL-22 and Malignant Cells

IL-22 promotes tumorigenesis by inducing cell proliferation, migration, angiogenesis, dysplasia, and oxidative stress [16]. The overexpression of IL-22 plays a very important role in many tumors, suggesting that IL-22 may become an important target for tumor treatment. However, the expression of IL-22 at different stages of tumor progression may have significantly different results. In the precancerous stage, overexpression of IL-22 is a physiological response that can effectively prevent the occurrence of tumors by enhancing host defense, eliminating bacteria and virus replication, and promoting tissue regeneration by IL-22, inducing proliferative

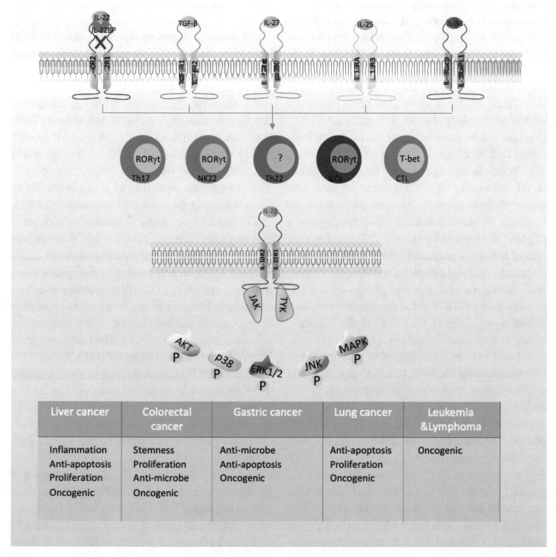

Fig. 5.1 The oncogenic roles of IL-22 within tumor microenvironment. IL-22 can be triggered by many stimulus within TME including IL-22, TGF-β, IL-27, IL-25, and IL-38 etc., mainly secreted from T helper and cytotoxic T cells as the response for the transcriptional factors such as RORγt and T-bet. Excessive IL-22 in turn promotes inflammation, proliferation, stemness, and anti-apoptosis by means of activation downstream signaling involving AKT, p38MAPK, ERK1/2, and JNK etc.

and antiapoptotic signaling pathways [16]. However, once a tumor is formed, the overexpression of IL-22 is a pathological response that has tumor-promoting effects by proliferative and antiapoptotic signaling pathways. However, the role of IL-22 is tissue-specific. Therefore, we summarize the roles of IL-22 in different malignancies.

5.3   Liver Cancer

Our research revealed that IL-22 expression was excessive in human hepatocellular carcinoma (HCC), and was mainly secreted by T cells and macrophages. HCC with poor differentiation has a stronger IL-22-containing TME [11]. Moreover, the high serum IL-22 levels in post-surgery HCC patients are significantly associated with poor prognosis [17]. Strong IL-23 and IL-22RA1 expression in human HCC provides evidence of both upstream and downstream signaling in an excessive IL-22-related TME. IL-22 has proliferation-promoting and antiapoptotic effects in different animal models. These two pro-tumor effects are mainly due to activation of STAT3 [11]. However, there is almost no evidence indicating that IL-22 has any effect on inflammation. This is due to strong evidence that liver-specific IL-22 transgenic (IL-22TG) mice develop normally without obvious adverse phenotypes or evidence of chronic inflammation (except for a slightly thicker epidermis and minor inflammation of the skin) compared to wild-type mice. In a classical diethylnitrosamine-induced mouse HCC model, two independent groups show a significant role of IL-22 in the development of liver cancer using either IL-22 KO or IL-22 TG mice, which is specifically dependent on the activation of STAT3 [11, 18]. IL-22 is also an ideal target for experimental HCC treatment. Oral metformin administration leads to a significant reduction in tumor growth, which is accompanied by decreased IL-22, IL-22-induced STAT3 phosphorylation, and inhibition of the upregulation of the downstream genes Bcl-2 and cyclin D1 [19]. At the cellular level, metformin attenuated Th1- and Th17-derived IL-22 production. Furthermore, metformin inhibits de novo generation of Th1 and Th17 cells from naive CD4+ cells [19]. In addition, STAT3 and IL-22 were revealed as multiple targets, and hepatocytes overexpressing IL-22 revealed a set of regulated antioxidants, mitogenic, and acute phase genes compared to wild-type mice. Hence, IL-22 blocked hepatic oxidative stress and its associated stress kinases via the induction of metallothionein, one of the most potent antioxidant proteins. Moreover, although it does not target immune cells, IL-22 treatment attenuates the inflammatory functions of hepatocyte-derived, mitochondrial DNA-enriched extracellular vesicles, thereby suppressing liver inflammation in nonalcoholic steatohepatitis, which is another cause of hepatocarcinogenesis [20, 21].

5.4 Colorectal Cancer

IL-22 was also significantly upregulated in both ulcerative colitis (UC) tissues and colon cancer tumor-infiltrated innate lymphoid cells. Moreover, our findings demonstrated that IL-22 expression was significantly higher in colon cancer tissues than in normal colon tissues. Both IL-22 receptor alpha 1 (IL-22RA1) and IL-23 are highly expressed in colorectal cancer (CC) and UC tissues compared to normal controls [12]. The downstream signaling of IL-22 is similar to other cancer types, mainly including STAT3, AKT, NF-κB, and MAPK kinases, which have antiapoptotic effects exerted by downstream genes, including B-cell lymphoma-extra large (Bcl-XL), Cyclin D1, deleted in malignant brain tumors 1 (DMBT1) [22], and vascular endothelial growth factor (VEGF). In addition to its antiapoptotic effects, IL-22 can also affect epigenetics of colon cancer stem cells by affecting the expression of the histone 3 lysine 79 (H3K79) methyltransferase DOT1L, and maintaining the stemness and tumorigenesis of cancer stem cells by upregulating NANOG, SOX2, and Pou5F1 [23]. IL-22 is also a metabolism-related gene, which enhances glucose consumption and lactate production via c-Myc- and STAT3-regulated hexokinase-2 [24]. Interestingly, IL-22 can be regulated by certain metabolites. Metabolites of glucosinolates upregulated IL-22 in type 3 innate lymphoid cells (ILC3) and gamma delta T cells via the aryl hydrocarbon receptor (AhR), which enhances the genome integrity of IL-22-producing cells and target cells, directly regulating components of the DNA damage response in epithelial stem cells [25]. IL-22 can also be triggered by intestinal microbes, such as *Helicobacter hepaticus* (*Hh*) [26] and *Lactobacillus* [27]. *Hh* enhances the production of IL-22 within crypt epithelial cells, which is associated with *Hh*-induced DNA damage and the development of dysplasia by promoting the production of inducible nitric oxide synthase [26]. *Lactobacillus* triggering IL-22 may potentially be a useful mucosal therapeutic agent for the treatment of graft versus host disease, provided that chromosomal integration of the IL-22 expression cassettes can be achieved [27].

5.5   Pancreatic Cancer (PC)

High levels of IL-22 have been detected in pancreatic tumor tissues and peripheral blood. IL-22 regulates the production of VEGF and the antiapoptotic factor Bcl-XL in IL-22R-positive PC cells [13]. In addition, IL-22 promotes immunosuppressive cytokines, including IL-10 and TGF-β1, thus downregulating intratumoral inflammation induced by natural killer (NK) cells [13, 28]. Other reports reveal that IL-22 originates from ILC3s in human PC, and IL-22R levels are increased in PC cells to ensure its downstream signaling, including STAT3 and AKT [29].

5.6   Gastric Cancer (GC)

IL-22 was reported to be associated with anti-*Helicobacter pylori* (*Hp*) responses in human GC. Synergistically with IL-17A, IL-22 enhances the anti-*Hp* ability of gastric epithelial cells (GECs), both in vitro and in vivo, by enhancing the production of antimicrobials and chemokines, such as IL-8, components of calprotectin (CP), lipocalin (LCN), and some beta-defensins in both human and primary mouse GECs and gastroids [30]. This report suggests that IL-22 is a protective cytokine in human GC development. However, the frequency of IL-22-positive T cells increased in tumor tissues compared to tumor-draining lymph nodes, non-tumor, and peritumoral tissues. Moreover, higher intratumoral IL-22+ CD4+ T-cell and Th22-cell percentages are found in patients with advanced tumor-node-metastasis stage and reduced overall survival [31]. Cancer-associated fibroblasts (CAFs) derived from IL-22 have also been reported in human GC, which can promote GC cell invasion via STAT3 and extracellular signal-regulated kinase (ERK) signaling [32]. Interestingly, IL-22 levels were elevated in the serum of elderly GC patients compared to healthy

elderly and young healthy controls. Peripheral IL-22 levels increase with age and are used as prognostic markers for identifying GC in elderly patients [33].

5.7   Lung Cancer

IL-22 is highly expressed in primary tumor tissue, malignant pleural effusion, and serum of patients with small- and large-cell lung carcinoma [34]. IL-22 also has an antiapoptotic role in human non-small cell lung cancers (NSCLCs), including serum starvation-induced and chemotherapeutic drug-induced apoptosis via activation of STAT3 and its downstream antiapoptotic proteins, such as Bcl-2 and Bcl-XL and inactivation of ERK 1/2. The internal signal transduction of IL-22 is also provided by overexpression of its receptor IL-22R1, both in human NSCLC cancer tissues and cell lines. However, the upregulation of IL-22R1 was associated with chemotherapy resistance to cisplatin-induced apoptosis, but not carboplatin-induced apoptosis, indicating different molecular mechanisms of chemotherapy resistance [35, 36]. Serum IL-22 combined with hepatocyte growth factor (HGF) and IL-20-inducible hepatocyte growth factor (iHGF) may be a prognostic factor for NSCLC progression before chemotherapy [37]. The cellular secretory sources of IL-22 in human lung cancer are varied, including myeloid, mixed T helper cell populations composed of Th1, Th17, and Th22 cells [38], and CAFs [39]. Naturally, the mechanism of IL-22 in human lung cancer also varies depending on its secretory source.

5.8   Breast Cancer

Using a mouse spontaneous breast cancer model, IL-22 was specifically upregulated in the TME during malignant transformation, and deletion of IL-22 blocked malignant transition, which reveals a role of IL-22 in the tumorigenesis of breast cancer [40]. IL-22 in breast cancer has been reported to be produced by the regulation of IL-1β/NOD-, LRR-, and pyrin domain-containing protein 3 (NLRP3) signaling; blockage of IL-1β by its antagonist anakinra abrogated IL-22 production and reduced tumor growth in a murine breast cancer model [38]. The important role of IL-22 has also been shown in clinical samples. The prevalence of IL-22-producing T cells gradually increases in normal, peritumoral, and tumor tissues [41]. Tumor IL-22 levels are associated with an aggressive phenotype in breast cancer [40]. Mechanistic studies indicate that IL-22 promotes human breast cancer through the Janus kinase (JAK)-STAT3/MAPKs/AKT [41], PI3K-AKT-mTOR [42], ERK1/2 [43], and c-Jun N-terminal kinase (JNK)/c-Jun [44] signaling pathways, promoting breast cancer cell proliferation and invasion. Administration of recombinant-IL-22 in the TME does not influence in vivo tumor initiation and proliferation, but only promotes malignant transformation of cancer cells [40].

5.9   Leukemia and Lymphoma

IL-22 and downstream signaling also play essential roles in the development and progression of leukemia or lymphoma. The IL-22-related autocrine stimulatory loop contributes to STAT3 activation and tumorigenicity of anaplastic large-cell lymphoma (ALCL) and anaplastic lymphoma kinase-positive (ALK (+)) ALCL. The IL-22 receptor is expressed in all ALK (+) ALCL cell lines and tumor tissues [45]. In addition, the fusion oncoprotein nucleophosmin (NPM)-ALK directly contributes to the aberrant expression of IL-22R1 and NPM-ALK overexpression in Jurkat cells with induced IL-22R1 expression and IL-22-mediated STAT3 activation [45]. Another study provided similar results in ALCL; IL-22 also originates from tumor cells, which is mediated by the tyrosine kinase 2 (TYK2)/STAT1/myeloid cell leukemia 1 (MCL1) axis [46]. Besides ALCL, IL-22 is also elevated in patients with clinical lymphocytic leukemia [47] and chronic myeloid leukemia [48], which suggests an oncogenic role of IL-22 in chronic leukemia. However, the function of IL-22 is varied in acute leukemia, overexpression of IL-22 is reported by two independent groups [49, 50], while other research indicates that Th22 cells and IL-22 are significantly decreased in newly diagnosed patients compared to complete-remission acute myeloid leukemia patients or controls [51].

5.10   IL-22 Is a Target for Drug Discovery

As mentioned above, IL-22 promotes cell proliferation, angiogenesis, and dysplasia, and its receptor, IL-22R1, is mainly distributed in epithelial or tumor cells. Therefore, antitumor therapy targeting the IL-22 signaling pathway may be a feasible approach [9, 16]. IL-22 neutralizing antibodies, such as ILV-094 have been registered for phase II clinical trials for psoriasis and rheumatoid arthritis (NCT01941537, http://clinicaltrials.gov/) [52, 53]. Neutralization of IL-22 might reduce metastasis, chemotherapy resistance, and tumor-related inflammation to effectively control tumor progression and improve the quality of life of patients with end-stage malignancies [9]. IL-22 binding protein (IL-22BP) is a special natural IL-22 antagonist; therefore, it might be an ideal target for anti-IL-22 therapy [54]. In addition, compared to specific anti-IL-22 drugs, some antitumor drugs targeting other molecules approved by the FDA can also effectively regulate the expression of IL-22 [9]. Treatment with tumor necrosis factor alpha (TNF-α) antibody, including adalimumab, etanercept, and infliximab temporarily reduced the expression of IL-22 because the differentiation of Th22 cells depends on the stimulation of TNF-α. The IL-6 antibody, tocilizumab, also inhibits Th17 and Th22 cell differentiation. In addition, a neutralizing antibody against IL-12p40, reduces the expression of IL-12 and IL-23; prevents the differentiation of Th1, Th17, and Th22 cells; and blocks the production of IL-22 [9]. However, it should be noted that blocking both TNF-α and IL-12p40 signaling pathways will cause serious side effects in the biological function of immune cells, comprehensively inhibit cytokines, and may facilitate tumor growth. STAT3 is an important part of the downstream signaling pathway of

IL-22, carrying out most tumor-promoting effects; therefore, a STAT3 inhibitor might be a reasonable adjuvant combination therapy of anti-IL-22 therapy. However, there are many problems. Because there is a tightly consistor monodule between STAT3 and STAT1, simply blocking STAT3 signaling might also disrupt STAT1-dependent apoptosis and innate immunity [55]. In general, any treatment regimen for IL-22 should be based on the principle of effective control of the tumor and minimization of extensive inflammatory responses and organ damage.

In summary, many studies have demonstrated that IL-22 promotes the development of various tumors. With regard to the exploration of IL-22 functions, balancing tissue repair and tumor-promoting roles of IL-22 might be a new direction.

References

1. Ouyang W et al (2011) Regulation and functions of the IL-10 family of cytokines in inflammation and disease. Annu Rev Immunol 29:71–109
2. Sheppard P et al (2003) IL-28, IL-29 and their class II cytokine receptor IL-28R. Nat Immunol 4(1):63–68
3. Kotenko SV (2002) The family of IL-10-related cytokines and their receptors: related, but to what extent? Cytokine Growth Factor Rev 13(3):223–240
4. Sabat R et al (2007) IL-19 and IL-20: two novel cytokines with importance in inflammatory diseases. Expert Opin Ther Targets 11(5):601–612
5. Wolk K, Sabat R (2006) Interleukin-22: a novel T- and NK-cell derived cytokine that regulates the biology of tissue cells. Cytokine Growth Factor Rev 17(5):367–380
6. Sanos SL et al (2009) RORgammat and commensal microflora are required for the differentiation of mucosal interleukin 22-producing NKp46+ cells. Nat Immunol 10(1):83–91
7. Takatori H et al (2009) Lymphoid tissue inducer-like cells are an innate source of IL-17 and IL-22. J Exp Med 206(1):35–41
8. Brand S et al (2007) IL-22-mediated liver cell regeneration is abrogated by SOCS-1/3 overexpression in vitro. Am J Physiol Gastrointest Liver Physiol 292(4):G1019–G1028
9. Sabat R, Ouyang W, Wolk K (2014) Therapeutic opportunities of the IL-22-IL-22R1 system. Nat Rev Drug Discov 13(1):21–38
10. Yu H, Pardoll D, Jove R (2009) STATs in cancer inflammation and immunity: a leading role for STAT3. Nat Rev Cancer 9(11):798–809
11. Jiang R et al (2011) Interleukin-22 promotes human hepatocellular carcinoma by activation of STAT3. Hepatology 54(3):900–909
12. Jiang R et al (2013) IL-22 is related to development of human colon cancer by activation of STAT3. BMC Cancer 13:59
13. Curd LM, Favors SE, Gregg RK (2012) Pro-tumour activity of interleukin-22 in HPAFII human pancreatic cancer cells. Clin Exp Immunol 168(2):192–199
14. Nardinocchi L et al (2015) Interleukin-17 and interleukin-22 promote tumor progression in human non-melanoma skin cancer. Eur J Immunol 45(3):922–931
15. Dudakov JA et al (2012) Interleukin-22 drives endogenous thymic regeneration in mice. Science 336(6077):91–95
16. Ouyang W, O'Garra A (2019) IL-10 family cytokines IL-10 and IL-22: from basic science to clinical translation. Immunity 50(4):871–891
17. Molina MF et al (2019) Type 3 cytokines in liver fibrosis and liver cancer. Cytokine 124:154497
18. Park O et al (2011) In vivo consequences of liver-specific interleukin-22 expression in mice: implications for human liver disease progression. Hepatology 54(1):252–261
19. Zhao D et al (2015) Metformin decreases IL-22 secretion to suppress tumor growth in an orthotopic mouse model of hepatocellular carcinoma. Int J Cancer 136(11):2556–2565
20. Hwang S et al (2019) Interleukin-22 ameliorates neutrophil-driven nonalcoholic steatohepatitis through multiple targets. Hepatology 72(2):412–429
21. Rolla S et al (2016) The balance between IL-17 and IL-22 produced by liver-infiltrating T-helper cells critically controls NASH development in mice. Clin Sci (Lond) 130(3):193–203
22. Fukui H et al (2011) DMBT1 is a novel gene induced by IL-22 in ulcerative colitis. Inflamm Bowel Dis 17(5):1177–1188
23. Kryczek I et al (2014) IL-22(+)CD4(+) T cells promote colorectal cancer stemness via STAT3 transcription factor activation and induction of the methyltransferase DOT1L. Immunity 40(5):772–784
24. Liu Y et al (2017) Interleukin-22 promotes aerobic glycolysis associated with tumor progression via targeting hexokinase-2 in human colon cancer cells. Oncotarget 8(15):25372–25383
25. Gronke K et al (2019) Interleukin-22 protects intestinal stem cells against genotoxic stress. Nature 566(7743):249–253
26. Wang C et al (2017) Interleukin-22 drives nitric oxide-dependent DNA damage and dysplasia in a murine model of colitis-associated cancer. Mucosal Immunol 10(6):1504–1517
27. Lin Y et al (2017) Lactobacillus delivery of bioactive interleukin-22. Microb Cell Factories 16(1):148
28. Xuan X et al (2018) Diverse effects of interleukin-22 on pancreatic diseases. Pancreatology 18(3):231–237
29. Xuan X et al (2020) ILC3 cells promote the proliferation and invasion of pancreatic cancer cells

through IL-22/AKT signaling. Clin Transl Oncol 22(4):563–575

30. Dixon BR et al (2016) IL-17a and IL-22 induce expression of antimicrobials in gastrointestinal epithelial cells and may contribute to epithelial cell defense against helicobacter pylori. PLoS One 11(2):e0148514

31. Zhuang Y et al (2012) Increased intratumoral IL-22-producing CD4(+) T cells and Th22 cells correlate with gastric cancer progression and predict poor patient survival. Cancer Immunol Immunother 61(11):1965–1975

32. Fukui H et al (2014) IL-22 produced by cancer-associated fibroblasts promotes gastric cancer cell invasion via STAT3 and ERK signaling. Br J Cancer 111(4):763–771

33. Chen X et al (2018) Accumulation of T-helper 22 cells, interleukin-22 and myeloid-derived suppressor cells promotes gastric cancer progression in elderly patients. Oncol Lett 16(1):253–261

34. Kobold S et al (2013) Interleukin-22 is frequently expressed in small- and large-cell lung cancer and promotes growth in chemotherapy-resistant cancer cells. J Thorac Oncol 8(8):1032–1042

35. Zhang W et al (2008) Antiapoptotic activity of autocrine interleukin-22 and therapeutic effects of interleukin-22-small interfering RNA on human lung cancer xenografts. Clin Cancer Res 14(20):6432–6439

36. Bi Y et al (2016) Interleukin-22 promotes lung cancer cell proliferation and migration via the IL-22R1/STAT3 and IL-22R1/AKT signaling pathways. Mol Cell Biochem 415(1-2):1–11

37. Naumnik W et al (2016) Clinical implications of hepatocyte growth factor, interleukin-20, and interleukin-22 in serum and bronchoalveolar fluid of patients with non-small cell lung cancer. Adv Exp Med Biol 952:41–49

38. Voigt C et al (2017) Cancer cells induce interleukin-22 production from memory CD4(+) T cells via interleukin-1 to promote tumor growth. Proc Natl Acad Sci U S A 114(49):12994–12999

39. Li H et al (2019) Interleukin-22 secreted by cancer-associated fibroblasts regulates the proliferation and metastasis of lung cancer cells via the PI3K-Akt-mTOR signaling pathway. Am J Transl Res 11(7):4077–4088

40. Katara GK et al (2020) Interleukin-22 promotes development of malignant lesions in a mouse model of spontaneous breast cancer. Mol Oncol 14(1):211–224

41. Wang S et al (2018) Interleukin-22 promotes triple negative breast cancer cells migration and paclitaxel

resistance through JAK-STAT3/MAPKs/AKT signaling pathways. Biochem Biophys Res Commun 503(3):1605–1609

42. Rui J et al (2017) IL-22 promotes the progression of breast cancer through regulating HOXB-AS5. Oncotarget 8(61):103601–103612

43. Weber GF et al (2006) IL-22-mediated tumor growth reduction correlates with inhibition of ERK1/2 and AKT phosphorylation and induction of cell cycle arrest in the G2-M phase. J Immunol 177(11):8266–8272

44. Kim K et al (2014) Interleukin-22 promotes epithelial cell transformation and breast tumorigenesis via MAP 3K8 activation. Carcinogenesis 35(6):1352–1361

45. Bard JD et al (2008) Aberrant expression of IL-22 receptor 1 and autocrine IL-22 stimulation contribute to tumorigenicity in ALK+ anaplastic large cell lymphoma. Leukemia 22(8):1595–1603

46. Prutsch N et al (2019) Dependency on the TYK2/STAT1/MCL1 axis in anaplastic large cell lymphoma. Leukemia 33(3):696–709

47. Kouzegaran S et al (2018) Elevated IL-17A and IL-22 regulate expression of inducible CD38 and Zap-70 in chronic lymphocytic leukemia. Cytometry B Clin Cytom 94(1):143–147

48. Chen P et al (2015) The alteration and clinical significance of Th22/Th17/Th1 cells in patients with chronic myeloid leukemia. J Immunol Res 2015:416123

49. Tian T et al (2013) Increased Th22 cells as well as Th17 cells in patients with adult T-cell acute lymphoblastic leukemia. Clin Chim Acta 426:108–113

50. Yu S et al (2014) Elevated Th22 cells correlated with Th17 cells in peripheral blood of patients with acute myeloid leukemia. Int J Mol Sci 15(2):1927–1945

51. Tian T et al (2015) The profile of T helper subsets in bone marrow microenvironment is distinct for different stages of acute myeloid leukemia patients and chemotherapy partly ameliorates these variations. PLoS One 10(7):e0131761

52. Jin M, Yoon J (2018) From bench to clinic: the potential of therapeutic targeting of the IL-22 signaling pathway in atopic dermatitis. Immune Netw 18(6):e42

53. Kragstrup TW et al (2018) The IL-20 cytokine family in rheumatoid arthritis and spondyloarthritis. Front Immunol 9:2226

54. Huber S et al (2012) IL-22BP is regulated by the inflammasome and modulates tumorigenesis in the intestine. Nature 491(7423):259–263

55. Fagard R et al (2013) STAT3 inhibitors for cancer therapy: have all roads been explored? JAKSTAT 2(1):e22882

IL-23 and the Tumor Microenvironment

6

Sweta Subhadarshani, Nabiha Yusuf, and Craig A. Elmets

Abstract

The tumor microenvironment (TME), which assists in the development, progression, and metastasis of malignant cells, is instrumental in virtually every step of tumor development. While a healthy TME can protect against malignancy, in an unhealthy state, it can result in aberrant cellular behavior and augment tumor progression. Cytokines are one component of the TME, therefore, understanding the composition of the cytokine milieu in the tumor microenvironment is critical to understand the biology of malignant transformation. One cytokine, interleukin (IL)-23, has received particular scrutiny in cancer research because of its ability to manipulate host immune responses, its role in modulating the cells in TME, and its capacity to directly affect a variety of premalignant and malignant tumors. IL-23 belongs to the IL-12 cytokine family, which is produced by activated dendritic cells (DC) and macrophages. IL-23 acts by binding to its receptor consisting of two distinct subunits, IL-12Rβ1 and IL-23R. This, in turn, leads to janus kinase (JAK) activation and signal transducer and activator of transcription (STAT) 3/4 phosphorylation. There have been contradictory reports of pro- and antitumor effects of IL-23, which likely depend on the genetic background, the type of tumor, the causative agent, and the critical balance of STAT3 signaling in both the tumor itself and the TME. Clinical trials of IL-12/23 inhibitors that are used to treat patients with psoriasis, have been scrutinized for reports of malignancy, the most common being nonmelanoma skin cancers (NMSCs). Continued investigation into the relationship of IL-23 and its downstream pathways holds promise in identifying novel targets for the management of cancer and other diseases.

Keywords

Tumor microenvironment · Interleukins · Tumor-associated macrophages · Angiogenesis · Matrix metallopeptidases · Pericytes · Melanocytic nevi · Nonmelanoma skin cancer · Melanoma · Multiple myeloma · Breast cancer · Colorectal cancer · Esophageal cancer · Pediatric B-acute lymphocytic leukemia · Psoriasis therapy

S. Subhadarshani · N. Yusuf · C. A. Elmets (✉)
Department of Dermatology, O'Neal Comprehensive Cancer Center, University of Alabama at Birmingham, University Boulevard, Birmingham, AL, USA
e-mail: celmets@uabmc.edu

© The Editor(s) (if applicable) and The Author(s), under exclusive license to Springer Nature Switzerland AG 2021
A. Birbrair (ed.), *Tumor Microenvironment*, Advances in Experimental Medicine and Biology 1290, https://doi.org/10.1007/978-3-030-55617-4_6

6.1   Introduction

The tumor microenvironment (TME), which assists in the development, progression, and metastasis of malignant cells, is comprised of cellular components and an extracellular matrix. The extracellular matrix (ECM) consists of collagen, elastin, proteoglycans, and hyaluronic acid. Fibroblasts, myofibroblasts, neuroendocrine (NE) cells, adipose cells, immune-inflammatory cells, and the blood and lymphatic vascular networks comprise the cellular component [1, 2]. The TME is instrumental in virtually every step of tumor pathogenesis. It is akin to the concept of "seed" and the "soil" in which the TME (soil) plays an important role in the growth and development of mutant cells (seed) and their evolution to clinically apparent malignancy and metastases. While a healthy TME can deter malignancy, and help protect against invasion in an unhealthy state, it can result in aberrant cellular behavior and progression to advanced malignancy.

Cytokines are one component of the TME. They are protein mediators controlling autocrine or paracrine communications involved in many biologic processes. The cytokine network controls innate and adaptive immune responses, cell growth, survival, inflammation, and differentiation. Cytokines play a myriad of positive and negative roles in tumorigenesis including regulation of leukocyte infiltration into tumors, stimulation of neovascularization and manipulation of the host immune response to tumor cells [2]. Therefore, understanding the composition of the cytokine milieu in the tumor microenvironment is fundamental to understanding of the biology of malignant transformation.

One cytokine, IL-23, has received particular scrutiny in cancer research because of its ability to manipulate host immune responses, its role in modulating the activities of cells and molecules in the tumor microenvironment, and its capacity to directly affect a variety of premalignant and malignant tumors. Interleukin-23 (IL-23) belongs to the IL-12 cytokine family, which includes IL-12, IL-23, IL-27, and IL-35 [3, 4]. It is a dimeric peptide molecule comprised of a p40 subunit, which it shares with IL-12, and a p19 subunit, which is unique to IL-23. Both IL-12 and IL-23 signal through heterodimeric receptors, with a common IL-12 receptor β1 (IL-12Rβ1), which is coupled with IL-12Rβ2 to form the IL-12 receptor and with IL-23R to form the IL-23 receptor (Fig. 6.1).

The IL-12 family of cytokines is produced by activated dendritic cells (DC) and macrophages that are stimulated by microbial and chemical pathogens, CD40L, and toll-like receptor (TLR) ligands [4]. IL-12 and IL-23 play important roles in the development of immune responses in various disease conditions (Fig. 6.2).

They act as a link between the innate and adaptive immune system. IL-23 contributes to the differentiation of naïve T cells into T-helper (Th)17 cells that produce IL-17; IL-12 facilitates the development of Th1 cells that produce interferon gamma (IFNγ). IL-12 and IL-23 also participate in the functions of different effector cell types. For example, IL-23 has been shown to suppress natural killer (NK) cell-mediated control of lung metastases by a perforin and IFNγ dependent mechanism. IL-23 is thought to suppress natural or cytokine-induced innate immunity, exerting a pro-tumor effect, independent of IL-17A. Downstream signaling occurs via the Janus kinase–signal transducers and activators of transcription (JAK-STAT) pathway. IL-23 involves both STAT3 and STAT4 signaling. In contrast, IL-12 utilizes STAT4 (Fig. 6.2).

The local balance between IL-12 and IL-23 has repeatedly been shown to play an important role in determining whether a pro- or antitumor immune response develops. The role of IL-12 in promoting antitumor immunity by mediating immune surveillance of tumors is well-recognized [5, 6]. On the other hand, there are contradictory reports of pro- and antitumor effects of IL-23 and its receptor IL-23R. Whether IL-23 acts in a pro- or anticarcinogenic manner may depend on the genetic background, the type of tumor, the cause (i.e., ultraviolet (UV) radiation, chemical, virus, etc.) and the critical balance of STAT3 signaling in both the tumor and the tumor cell microenvironment. For example, in mice subjected to a two-stage skin tumorigenesis protocol using

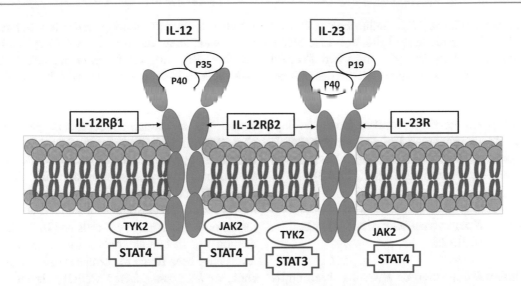

Fig. 6.1 IL-12 and IL-23 cytokines. IL-12 is composed of IL-12p40 and p35 subunits that bind to IL-12Rβ1 and IL-12Rβ2, respectively. IL-23 is a heterodimeric cytokine composed of p40 and p19 submits that bind to the IL-12βR1 and the IL-23 receptor (IL-23R), respectively. IL-12Rβ1 binds the JAK family member tyrosine kinase (TYK)2, whereas IL-12Rβ2 and IL-23R associate with JAK2. IL-23 stimulation activates the JAKs and STATs, but STAT3 and not STAT4, appears to be the predominant STAT-activated. *IL* interleukin, *JAK* Janus kinase, *TYK2* tyrosine kinase 2, *STAT* signal transducer and activator of transcription

Fig. 6.2 IL-12 and IL-23 activate the innate and adaptive immune responses. Activated DC and macrophages that are stimulated by microbial pathogens and chemicals, CD40L, and TLR ligands produce the IL-12 family of cytokines. IL-23 contributes to the differentiation of naïve T cells into Th17 cells that produce IL-17; IL-12 facilitates the development of Th1 cells that produce IFNγ. IL-12 and IL-23 also participate in the functions of different cell types. *DC* dendritic cells, *TLR* toll-like receptors, *Th* T-helper, *IFN* interferon, *TNF* tumor necrosis factor

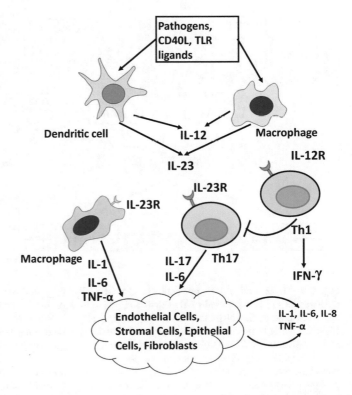

7,12-dimethylbenz (a) anthracene (DMBA) and 12-*O*-tetradecanoylphorbol-13-acetate (TPA), IL-23 knockout (KO) mice were found to be resistant to skin tumor development, and IL-12-mediated immunotherapy was more effective in the absence of IL-23p19 [7]. On the other hand, IL-23 deficiency blocked tumor development in a murine model of photocarcinogenesis despite the fact that it enhanced keratinocyte tumors in the same animals [8] (Fig. 6.3).

6.2 Non-immunological Effects of IL-23

While most attention on IL-23 has been on its role to facilitate activation of the Th17/IL-17 pathway, IL-23 has a number of other activities,

which may impact tumorigenesis. It interferes with the antitumor function of NK cells by blocking the IFNγ and perforin-mediated effects. It also supports neoangiogenesis and inhibits CD8 T-cell infiltration into the tumor tissue. IL-23 also activates DNA repair pathways, an activity which occurs via an immune-independent pathway. Under physiologic situations, IL-23 favors high bone mass by reducing bone resorption. In contrast, in pathological circumstances, it has a stimulatory effect on osteoclast formation, mainly via the induction of receptor activator of nuclear factor kappa-B (RANKL) by T cells and IL-17 production [9]. In human studies, increased IL-23 levels have been found to correlate with vascular endothelial growth factor (VEGF) levels in colorectal carcinoma [10]. Sheng et al. found that IL-23 signaling upregulated proteins that were

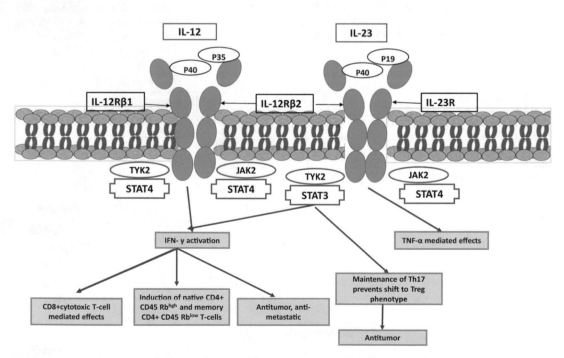

Fig. 6.3 IL-12 and IL-23 activate the JAK-STAT pathway. IL-12 is composed of IL-12p40 and p35 subunits that bind to IL-12Rβ1 and IL-12Rβ2, respectively. Ligand binding brings the receptor chains and associated JAKs (JAK2 and TYK2) in close proximity, resulting in JAK transphosphorylation and subsequent phosphorylation of the receptor chains by activated JAKs. IL-12Rβ2 is phosphorylated and serves as a docking site for STAT4. STAT4 binds to the receptor chain and is itself phosphorylated. STAT4 homodimers shuttle into the nucleus where

they bind to STAT-binding sites in the interferon (IFN)-γ promoter, and induce transcription of the IFN-γ gene. IFN-γ activation can activate CD8+ cytotoxic T cells, induce naïve and memory T cells, and can inhibit tumor growth and metastasis. IL-23 can have pro- or anticarcinogenic effects depending on the critical balance of STAT3 signaling in the tumor and tumor cell microenvironment. *IL* interleukin, *JAK* Janus kinase, *TYK2* tyrosine kinase 2, *STAT* signal transducer and activator of transcription, *IFN* interferon correlated with proliferation and survival of breast cancer cells [11]. IL-23 also promotes tumor angiogenesis [12]. It is reasonable to speculate that IL-23 has an effect on type-2 pericytes, which have an important role in tumor angiogenesis as well [13].

6.3 Keratinocyte Carcinomas

Langowski et al. were the first to study the role of IL-23 in carcinogenesis using the mouse two-step skin carcinogenesis model, in which mice were initially treated with a single application of the carcinogen DMBA followed by repeated exposure of the treatment area to TPA [14]. When experiments were conducted in IL-23p19-deficient mice, they found an increase in infiltrating CD8+ T cells and reduced levels of IL-17, matrix metallopeptidase 9 (MMP9) and CD31 expression in carcinogen-treated skin compared to controls. This coincided with a reduction in cutaneous tumors compared to wild-type mice treated in the same manner. They concluded that IL-23 promoted the development of skin cancer by inducing expression of MMP9 and other genes involved in angiogenesis while reducing the infiltration of CD8+ T cells into skin tumors [14]. (Fig. 6.4).

In contrast to that seen in chemical skin carcinogenesis, in a murine model of photocarcinogenesis, IL-23p19 KO mice had an increase in UV-induced tumors. In this study, C57BL/6 mice lacking either p35 (IL-12p35) or p19 (IL-23p19) were subjected to a chronic UV photocarcinogenesis protocol. Mice lacking IL-23p19 developed significantly more tumors than wild-type mice. The opposite was observed in IL-12p35 KO mice. This study concluded that loss of IL-23, but not of IL-12, enhances the development of UVR-induced skin tumors, indicating that IL-23, but not IL-12, may counteract photocarcinogenesis. They also observed that non-epithelial tumors developed significantly earlier in IL-23 KO mice than in controls.

UVB-induced cyclobutane pyrimidine dimer (CPD) photoproducts cause mutations in tumor suppressor genes leading to photocarcinogenesis.

CPDs are removed by nucleotide excision repair (NER). Both IL-12 and IL-23 are capable of promoting removal of CPD by NER. In an experiment with p19, p35, and p40 KO mice, an increased risk of UV-induced skin cancer was seen in IL-23 p19 and IL-12/23p40 KO mice but not p35 KO mice, which suggest that loss of IL-12 seems to be compensated by IL-23 but not vice versa.

6.4 Melanoma

Nasti et al. [15] studied mice deficient in IL-23p19, IL-12p35, and IL-12/23 p40 using a two-stage melanoma genesis protocol they had developed in which DMBA and TPA were applied to the skin of C3H/HeN mice [15]. They found that mice deficient in IL-23p19 developed 70% more nevi, which grew 40% larger than the nevi of wild-type mice. Surprisingly, IL-23p35 KO mice displayed a reduction in the number and growth of nevi. They also observed that melanocytic cell lines established from IL-12p35 KO mice possessed fewer H-ras mutations compared with cell lines derived from IL 23p19 KO and wild-type mice. In that study, the level of nuclear DNA damage post-DMBA treatment was studied using gamma H2AX red immunofluorescence. They observed that DNA repair was augmented in IL-12p35 KO mice. In contrast, IL-23p19 KO mice demonstrated high positivity for gamma H2AX indicating poor DNA repair. Also, addition of recombinant IL-23 in the culture medium promoted DNA repair in melanocytes.

6.5 Breast Cancer

Sheng et al. [11] studied the IL-23/IL-23R axis in breast cancer patients and found that IL-23-mediated responses were crucial for tumor progression [11]. IL-23/IL-23R gene expression levels were markedly higher in tumors than in adjacent tissues and showed a positive correlation with patients' tumor size, TNM stage and metastasis, suggesting that IL-23 might be a potential prognostic marker and treatment target.

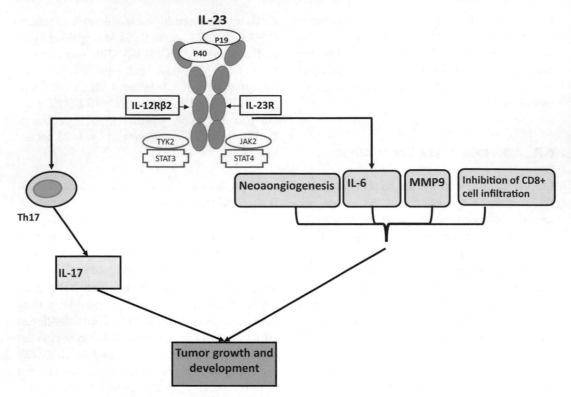

Fig. 6.4 IL-23 has an impact on tumorigenesis. IL-23 interferes with the antitumor function of NK cells by blocking the IFNγ and perforin-mediated effects. It also supports neoangiogenesis and inhibits CD8 T-cell infiltration into the tumor tissue. IL-23 has a stimulatory effect on IL-17 production. IL-23 also promotes tumor angiogenesis. *IL* interleukin, *JAK* Janus kinase, *TYK2* tyrosine kinase 2, *STAT* signal transducer and activator of transcription, *MMP* matrix metalloproteinase

The JAK2/STAT3 signal transduction pathway are downstream mediators of IL-23 signaling. JAK2/STAT3 activation was associated with poorer outcomes in metastatic breast cancer patients. In other studies, IL-23 inhibited apoptosis of breast cancer cells [16]. However, there was no significant difference in IL-23 levels among patients based on biomolecular characteristics, the different subtypes, or the presence of metastatic disease.

Shino et al. reported a case of breast cancer metastasis in the skin with increased levels of IL-23 and IL-17 [17]. They found that IL-23 and chemokine (C–C) ligand 20 (CCL20) were produced by tumor cells. IL-17-producing cells and CD163+ tumor-associated macrophages (TAMs) were found microscopically around tumor nodules consistent with the concept that IL-23

orchestrates an immunosuppressive tumor microenvironment at the site of breast cancer metastases. In other studies, evaluating the role of IL-23 in breast cancer, it has been found to facilitate angiogenesis, production of immunosuppressive cytokines, and infiltration of M2 macrophages and neutrophils, while at the same time suppressing antitumor immune responses through a reduction of CD4+ and CD8+ T cells [18].

6.6 Multiple Myeloma

The IL-23 receptor is present on normal plasma cells and on multiple myeloma (MM) cells [9]. Increased levels of IL-23 have been observed in the serum and bone marrow of MM patients compared to healthy controls. However, it does not appear to influence the proliferation of myeloma cells, nor does it stimulate apoptosis, inflammatory cell chemotaxis, or angiogenesis.

Kasamatu et al. observed that the IL-23R HH genotype was significantly associated with poor survival compared with the QH and HH genotypes in multiple myeloma patients [19].

6.7   Pediatric B-Acute Lymphocytic Leukemia (B-ALL)

IL-23 has been found to have antitumor actions in pediatric B-ALL [20]. In vitro studies and animal models have shown that IL-23 inhibits proliferation and augments apoptosis of malignant cells. This occurs through downregulation of cyclin D1 and Bcl-2. Interestingly, the effect is mediated by mir-15a [21].

6.8   Colorectal Cancer

A variety of studies implicate IL-23 and its receptor in colorectal cancer (CRC) growth and development [10, 19, 22, 23]. There is a significant increase in the serum level of the cytokine IL-23 in CRC patients as compared to healthy controls [10, 19]. Research has also shown a highly significant increase of serum levels of IL-23 with advanced TNM stages of CRC [19]. Increased levels of IL-23 have been found to correlate with the expression of VEGF and histological grades [10]. In addition, IL 23 has been found to play a pivotal role in the pathogenesis of inflammatory bowel disease (IBD) and colitis-associated colon cancer [22].

Basic leucine zipper ATF-like transcription factor (BATF) is a transcription factor that in ulcerative colitis, but not Crohn's disease, is elevated and the level of increase correlates with the levels of IL-23. In CRC, BATF correlates with both with IL-23 and IL-23R [23]. This has been investigated further in animal models in which BATF was deleted. Those animals had low levels of IL-23. Since BATF is a transcription factor for Th17 differentiation, the findings have

been interpreted as suggesting that BATF decreases Th17 cell differentiation which in turn leads to reduced levels of IL-23. Suzuki et al. studied the IL-23R expression in human colorectal cancer tissue samples. They found that all of the TNM stage IV patients were positive for IL-23R and that IL-23R was relatively high at the deepest point of invasion in some cases. In tissue culture, the proliferative and invasive activities and/or transforming growth factor beta (TGF-β) production were increased with IL-23 stimulation. This suggests that an autocrine mechanism via TGF-β has a pro-tumorigenic effect on CRC. IL-23 may therefore be a potential target for CRC immunotherapy.

In contrast to the pro-tumorigenic effect of IL-23 in colon cancer, some studies have found that IL-23 has an antitumor and antimetastatic role in this disease. In 2006, Shan et al. suggested that there is role for Th1 and dendritic cells in IL-23-mediated antitumor activity in murine adenocarcinoma cells [24]. The effect of IL-23 on tumor growth was also evaluated by Lo et al. in 2003 [25]. They observed that CD8+ T cells, but not CD4+ T cells or NK cells, were crucial for the antitumor activity of IL-23. In their study, murine colon adenocarcinoma cells were transduced with vectors so that they released single-chain IL-23 (scIL-23). The transduced cells were observed to have significantly greater antitumor activity than empty vector-treated cells. The antitumor effect was mediated by CD8+ T cells; neither CD4+ T cells nor NK cells were required. Wang et al. confirmed the role of cytotoxic T cells in the antitumor effects in an IL-23 transduced colon cancer cell model and went on to show that the antitumor effects in this model system were mediated by interferon-gamma producing CD8+ T cells [26].

6.9   Esophageal Cancer

There is evidence that IL-23 has a deleterious effect on esophageal cancer. When esophageal tissue was stained for IL-23 expression by immunohistochemistry, increased areas of IL-23 staining could be found in tissue from primary tumors of metastatic patients compared to primary tumors of nonmetastatic cancers or the precancerous lesions [27]. Incubation of esophageal carcinoma cell lines with IL-23 caused an increase in biomarkers associated with epithelial-mesenchymal transition as well as MMP9 and VEGF. These effects were mediated by IL-23 actions on the Wnt/beta-catenin pathway [27]. Consistent with a role for IL-23 in esophageal cancer, Chu et al. reported two potentially functional genetic single-nucleotide polymorphisms in IL-23R (SNPs; IL-23R rs6682925 T > C and rs1884444 T > G), which increase the risk of esophageal cancer [28].

6.10 Implication for Anti-IL-23 in Therapeutics

6.10.1 Anti IL-12/23p40

Ustekinumab and briakinumab (anti IL-12/23p40) have been used in treatments of psoriasis, psoriatic arthritis, inflammatory bowel disease and other immune-mediated conditions. Ustekinumab is approved by the Food and Drug Administration (FDA); briakinumab development was discontinued prior to FDA approval. Clinical trials of IL-12/23 inhibitors have been scrutinized for reports of malignancy, the most common being nonmelanoma skin cancers (NMSCs).

The prescribing information (PI) for ustekinumab contains a general warning that it "may increase risk of malignancy." This is based on the following observations from post-marketing safety data: (1) among patients treated with ustekinumab (3.2 years' median follow-up), Non-melanoma skin cancers (NMSCs) were reported in 1.5% of patients and malignancies excluding NMSCs were reported in 1.7% of patients; (2) aside from NMSCs, the most frequently observed malignancies were prostate, melanoma, colorectal, and breast. However, the profile was similar to the general population as adjusted for age, gender, and race; and (3) in post-marketing surveillance data of patients taking ustekinumab, there have been reports of rapid appearance of multiple cutaneous squamous cell carcinomas (SCCs) in those who had preexisting risk factors for NMSC.(Janssen Biotech Inc. Stelara® (ustekinumab) prescribing information, https://www.stelarainfo.com/pdf/prescribinginformation.) In a pooled analysis, the risk of SCC was similar to the risk of basal cell carcinoma (BCC) in patients treated with briakinumab, but patients were only taking the medication for relatively short periods of time [29]. In trials of ustekinumab for psoriasis, the proportion of BCCs was higher than that of SCCs which is similar to the general population [30].

6.10.2 Anti-IL-23p19

Several different IL-23p19-specific antibodies have obtained Food and Drug Administration (FDA) approval for moderate-to-severe psoriasis. Guselkumab was the first IL-23p19 antibody to be approved. Recently, the IL-23-specific inhibitors tildrakizumab and risankizumab have also been approved, and mirikizumab (LY3074828) has completed phase 2 studies. All have shown excellent efficacy in plaque psoriasis. There is limited data on malignancy risk, although NMSCs have been reported in some trials [31, 32].

6.11 Conclusions

IL-23 acts by binding to its receptor consisting of two distinct subunits, IL-12Rβ1 and IL-23R. This, in turn, leads to JAK activation and STAT3/4 phosphorylation. Although various studies in the last decade have implicated IL-23 in tumorigenesis of various tissues, its exact mechanism in the TME is still elusive. There have been contradictory reports of pro- and antitumor effects of IL-23, which likely depend on the genetic background, the type of tumor, the causative agent, and the critical balance of STAT3 signaling in both the tumor itself and the TME. The pro-tumor functions may involve a downstream IL-17-mediated signaling or act via an IL-17-independent pathway. The latter includes promotion of VEGF-mediated neoangiogenesis and MMP9 activation, inhibition of apoptosis and modulation of NK and T-cell function. Continued investigation into the relationship of IL-23 and its downstream pathways holds promise in identifying novel targets for cancer therapy.

References

1. Wang M, Zhao J, Zhang L, Wei F, Lian Y et al (2017) Role of tumor microenvironment in tumorigenesis. J Cancer 8(5):761–773. https://doi.org/10.7150/jca.17648
2. Wilson J, Balkwill F (2002) The role of cytokines in the epithelial cancer microenvironment. Semin Cancer Biol 12(2):113–120. https://doi.org/10.1006/scbi.2001.0419
3. Croxford AL, Mair F, Becher B (2012) IL-23: one cytokine in control of autoimmunity. Eur J Immunol 42(9):2263–2273. https://doi.org/10.1002/eji.201242598
4. Vignali DA, Kuchroo VK (2012) IL-12 family cytokines: immunological playmakers. Nat Immunol 13(8):722–728. https://doi.org/10.1038/ni.2366
5. Engel MA, Neurath MF (2010) Anticancer properties of the IL-12 family—focus on colorectal cancer. Curr Med Chem 17(29):3303–3308. https://doi.org/10.2174/092986710793176366
6. Tugues S, Burkhard SH, Ohs I, Vrohlings M, Nussbaum K et al (2015) New insights into IL-12-mediated tumor suppression. Cell Death Differ 22(2):237–246. https://doi.org/10.1038/cdd.2014.134
7. Teng MW, Andrews DM, McLaughlin N, von Scheidt B, Ngiow SF et al (2010) IL-23 suppresses innate immune response independently of IL-17A during carcinogenesis and metastasis. Proc Natl Acad Sci U S A 107(18):8328–8333. https://doi.org/10.1073/pnas.1003251107
8. Jantschitsch C, Weichenthal M, Proksch E, Schwarz T, Schwarz A (2012) IL-12 and IL-23 affect photocarcinogenesis differently. J Invest Dermatol 132(5):1479–1486. https://doi.org/10.1038/jid.2011.469
9. Giuliani N, Airoldi I (2011) Novel insights into the role of interleukin-27 and interleukin-23 in human malignant and normal plasma cells. Clin Cancer Res 17(22):6963–6970. https://doi.org/10.1158/1078-0432.CCR-11-1724
10. Ljujic B, Radosavljevic G, Jovanovic I, Pavlovic S, Zdravkovic N et al (2010) Elevated serum level of IL-23 correlates with expression of VEGF in human colorectal carcinoma. Arch Med Res 41(3):182–189. https://doi.org/10.1016/j.arcmed.2010.02.009
11. Sheng S, Zhang J, Ai J, Hao X, Luan R (2018) Aberrant expression of IL-23/IL-23R in patients with breast cancer and its clinical significance. Mol Med Rep 17(3):4639–4644. https://doi.org/10.3892/mmr.2018.8427
12. Langowski JL, Kastelein RA, Oft M (2007) Swords into plowshares: IL-23 repurposes tumor immune surveillance. Trends Immunol 28(5):207–212
13. Dulmovits A, Fleury T, White FM et al (2014) Time B. pericytes participate in normal and tumoral angiogenesis. Am J Physiol Cell Physiol 307(1):C25–C38
14. Langowski JL, Zhang X, Wu L, Mattson JD, Chen T et al (2006) IL-23 promotes tumour incidence and growth. Nature 442(7101):461–465. https://doi.org/10.1038/nature04808
15. Nasti TH, Cochran JB, Vachhani RV, McKay K, Tsuruta Y et al (2017) IL-23 inhibits melanoma development by augmenting DNA repair and modulating T cell subpopulations. J Immunol 198(2):950–961. https://doi.org/10.4049/jimmunol.1601455
16. Gangemi S, Minciullo P, Adamo B, Franchina T, Ricciardi GR et al (2012) Clinical significance of circulating interleukin-23 as a prognostic factor in breast cancer patients. J Cell Biochem 113(6):2122–2125. https://doi.org/10.1002/jcb.24083
17. Ishihara-Yusa S, Fujimura T, Lyu C, Sugawara M, Sakamoto K et al (2018) Breast cancer metastasis in the skin with hyperkeratotic pigmentation caused by melanocyte colonization. Case Rep Oncol 11(3):660–664. https://doi.org/10.1159/000493186
18. Nie W, Yu T, Sang Y, Gao X (2017) Tumor-promoting effect of IL-23 in mammary cancer mediated by infiltration of M2 macrophages and neutrophils in tumor microenvironment. Biochem Biophys Res Commun 482(4):1400–1406. https://doi.org/10.1016/j.bbrc.2016.12.048
19. Elessawi DF, Alkady MM, Ibrahim IM (2019) Diagnostic and prognostic value of serum IL-23 in colorectal cancer. Arab J Gastroenterol 20(2):65–68. https://doi.org/10.1016/j.ajg.2019.05.002
20. Cocco C, Canale S, Frasson C, Di Carlo E, Ognio E et al (2010) Interleukin-23 acts as antitumor agent on childhood B-acute lymphoblastic leukemia cells. Blood 116(19):3887–3898. https://doi.org/10.1182/blood-2009-10-248245
21. Cimmino A, Calin GA, Fabbri M, Iorio MV, Ferracin M et al (2005) miR-15 and miR-16 induce apoptosis by targeting BCL2. Proc Natl Acad Sci U S A 102(39):13944–13949. https://doi.org/10.1073/pnas.0506654102
22. Neurath MF (2019) IL-23 in inflammatory bowel diseases and colon cancer. Cytokine Growth Factor Rev 45:1–8. https://doi.org/10.1016/j.cytogfr.2018.12.002
23. Punkenburg E, Vogler T, Buttner M, Amann K, Waldner M et al (2016) Batf-dependent Th17 cells critically regulate IL-23 driven colitis-associated colon cancer. Gut 65(7):1139–1150. https://doi.org/10.1136/gutjnl-2014-308227
24. Shan BE, Hao JS, Li QX, Tagawa M (2006) Antitumor activity and immune enhancement of murine interleukin-23 expressed in murine colon carcinoma cells. Cell Mol Immunol 3(1):47–52. https://www.ncbi.nlm.nih.gov/pubmed/16549049

25. Lo CH, Lee SC, Wu PY, Pan WY, Su J et al (2003) Antitumor and antimetastatic activity of IL-23. J Immunol 171(2):600–607. https://doi.org/10.4049/jimmunol.171.2.600

26. Wang YQ, Ugai S, Shimozato O, Yu L, Kawamura K et al (2003) Induction of systemic immunity by expression of interleukin-23 in murine colon carcinoma cells. Int J Cancer 105(6):820–824. https://doi.org/10.1002/ijc.11160

27. Chen D, Li W, Liu S, Su Y, Han G et al (2015) Interleukin-23 promotes the epithelial-mesenchymal transition of oesophageal carcinoma cells via the Wnt/beta-catenin pathway. Sci Rep 5:8604. https://doi.org/10.1038/srep08604

28. Chu H, Cao W, Chen W, Pan S, Xiao Y et al (2012) Potentially functional polymorphisms in IL-23 receptor and risk of esophageal cancer in a Chinese population. Int J Cancer 130(5):1093–1097. https://doi.org/10.1002/ijc.26130

29. Langley RG, Papp K, Gottlieb AB, Krueger GG, Gordon KB et al (2013) Safety results from a pooled analysis of randomized, controlled phase II and III clinical trials and interim data from an open-label extension trial of the interleukin-12/23 monoclonal antibody, briakinumab, in moderate to severe psoriasis. J Eur Acad Dermatol Venereol 27(10):1252–1261. https://doi.org/10.1111/j.1468-3083.2012.04705.x

30. Pouplard C, Brenaut E, Horreau C, Barnetche T, Misery L et al (2013) Risk of cancer in psoriasis: a systematic review and meta-analysis of epidemiological studies. J Eur Acad Dermatol Venereol 27(Suppl 3):36–46. https://doi.org/10.1111/jdv.12165

31. Papp KA, Blauvelt A, Bukhalo M, Gooderham M, Krueger JG et al (2017) Risankizumab versus ustekinumab for moderate-to-severe plaque psoriasis. N Engl J Med 376(16):1551–1560. https://doi.org/10.1056/NEJMoa1607017

32. Reich K, Papp KA, Blauvelt A, Tyring SK, Sinclair R et al (2017) Tildrakizumab versus placebo or etanercept for chronic plaque psoriasis (reSURFACE 1 and reSURFACE 2): results from two randomised controlled, phase 3 trials. Lancet 390(10091):276–288. https://doi.org/10.1016/S0140-6736(17)31279-5

Interleukin (IL)-24: Reconfiguring the Tumor Microenvironment for Eliciting Antitumor Response

7

Rajagopal Ramesh, Rebaz Ahmed, and Anupama Munshi

Abstract

Interleukin (IL)-24 is a member of the IL-10 family of cytokines. Due to its unique ability to function as both a tumor suppressor and cytokine, IL-24-based cancer therapy has been developed for treating a broad spectrum of human cancers. Majority of the studies reported to date have focused on establishing IL-24 as a cancer therapeutic by primarily focusing on tumor cell killing. However, the ability of IL-24 treatment on modulating the tumor microenvironment and immune response is underinvestigated. In this article, we summarize the biological and functional properties of IL-24 and the benefits of applying IL-24-based therapy for cancer.

Keywords

Angiogenesis · Apoptosis · Cancer · Cancer stem cells · Cytokines · Glycosylation · IL-10 · IL-19 · IL-20 receptor · IL-22 receptor · IL-24 · Immunity · mda-7 · Metastasis · Phosphorylation · Post-translational modification · Ubiquitination

R. Ramesh (✉)
Department of Pathology, University of Oklahoma Health Sciences Center, Oklahoma City, OK, USA

Stephenson Cancer Center, University of Oklahoma Health Sciences Center, Oklahoma City, OK, USA

Graduate Program in Biomedical Sciences, Oklahoma City, OK, USA
e-mail: rajagopal-ramesh@ouhsc.edu

R. Ahmed
Department of Pathology, University of Oklahoma Health Sciences Center, Oklahoma City, OK, USA

Graduate Program in Biomedical Sciences, Oklahoma City, OK, USA
e-mail: rebaz-ahmed@ouhsc.edu

A. Munshi
Stephenson Cancer Center, University of Oklahoma Health Sciences Center, Oklahoma City, OK, USA

Department of Radiation Oncology, University of Oklahoma Health Sciences Center, Oklahoma City, OK, USA
e-mail: anupama-munshi@ouhsc.edu

Abbreviations

Ad	Adenovirus
CSC	Cancer stem cells
HDAC	Histone deacetylase
HMVEC	Human microvascular endothelial cells
Hsp-90	Heat shock protein-90
HUVEC	Human umbilical vein endothelial cells
i.t.	Intratumoral
i.v.	Intravenous

A. Birbrair (ed.), *Tumor Microenvironment*, Advances in Experimental Medicine and Biology 1290,
https://doi.org/10.1007/978-3-030-55617-4_7

IFN-γ	Interferon gamma
IL-10	Interleukin-10
IL-20R	Interleukin-20 receptor
IL-22R	Interleukin-22 receptor
IL-24	Interleukin-24
IP-10	Interferon gamma inducible 10
MDA-7	Melanoma differentiation-associated (MDA) gene 7
PERK	Protein kinase R-like endoplasmic reticulum kinase
PKR	Protein kinase R
PTM	Post-translational modification
siRNA	Small interfering RNA
STAT-3	Signal transducer and activator of transcription 3
TME	Tumor microenvironment
TSG	Tumor suppressor gene
VEGF	Vascular endothelial growth factor

7.1   Introduction

Interleukin-(IL)-24, previously referred to as melanoma differentiation-associated gene-(MDA)-7, is a member of the IL-10 cytokine family [1–3]. Its family members include IL-10, IL-19, IL-20, IL-22, and IL-26 [4–7]. IL-24 is located on chromosome 1 at 1q32 [3]. The IL-24 cDNA encodes an evolutionarily conserved protein of 206 amino acids with a predicted size of 23.8 kDa [8, 9]. IL-24 protein sequence predicted three glycosylation sites and five phosphorylation sites [10, 11]. Further, the presence of a secretory signal sequence in the cDNA enables the secretion of IL-24 protein and its ability to operate intracellularly and extracellularly in an autocrine and paracrine fashion [12, 13]. Finally, ubiquitination of IL-24 protein facilitating its intracellular half-life and function has been reported using lung tumor model [14]. The results from these studies indicated that IL-24 is susceptible to post-translational modification (PTM) and that the nature of PTM likely dictates the function of IL-24 to operate either as a tumor suppressor gene (TSG) or as a cytokine. Molecular and biochemical studies subsequently reported the identification of receptors for IL-24.

Studies showed IL-24 utilized two heterodimer receptor complexes, namely IL-20R1/IL-20R2 and IL-22R1/IL-20R2 complex [15, 16]. mRNA and protein expression analysis showed IL-24 mRNA but not the protein is detected in human cancer cell lines and tissues. Further, occurrence of IL-24 gene mutation or polymorphism has not been reported to date. Thus, the mechanism of IL-24 regulation in various cell types especially in human cancer cells remains to be elucidated.

The unique and distinct features of IL-24 separate it from other family members that has allowed its testing as a cancer therapeutic against a broad spectrum of human cancers both in the laboratory and in the clinic. Herein, we discuss the tumor suppressor and cytokine properties of IL-24 in reconfiguring the tumor microenvironment (TME) and the benefits in testing IL-24 as a cancer therapeutic.

7.2   IL-24 as a Tumor Suppressor and Inducer of Cell Death

The first report on IL-24 (mda-7) was made by Jiang et al. [8]. In that study, the authors identified induction of mda-7 expression in human melanoma cells made to differentiate with interferon gamma. Expression of mda-7 resulted in differentiation and cell cycle arrest. Based on this study, IL-24 was tested as a therapeutic in a breast cancer model. Overexpression of IL-24 resulted in induction of apoptosis and inhibition of tumor cell growth both in vitro and in vivo [17]. Analysis for IL-24 protein expression in human melanoma tissue samples showed its expression was detected in nevi and primary melanoma and progressively decreased with disease stage and was completely lost in metastatic melanoma [18]. In human lung cancer, IL-24 protein expression was detected in primary tumor tissues and loss of its expression correlated with poor clinical outcome [19]. Delivery of IL-24 using an adenoviral vector (Ad-mda7) in human lung tumor cell lines, in vitro, induced G1 phase cell-cycle arrest and apoptosis thereby resulting in suppression of tumor cell proliferation [20]. Furthermore, inhibition of tumor cell migration and invasion was observed in Ad-mda7-treated tumor cell lines. Finally, Ad-mda7 killed tumor cells but not normal cells demonstrating tumor cell selectivity, a feature preferred in cancer therapy. In vivo studies showed intratumoral (i.t.) administration of Ad-mda7 into lung tumors established in nude mice significantly suppressed tumor growth that was accompanied by apoptosis and reduced tumor vasculature as evidenced by reduction in CD31 positive staining [21]. In the same study, Ad-mda7 was shown to inhibit endothelial tube formation, a phenomenon reflective of potential antiangiogenic activity. Based on this observation, the authors of the study speculated for the first time that Ad-mda7 apart from having antitumor activity likely has antiangiogenic activity as well. All of these studies clearly established IL-24 as a TSG that resulted in testing of IL-24 as a cancer gene therapeutic in several laboratories.

Introduction of IL-24 using viral and nonviral vectors in a broad spectrum of human cancer cell lines resulted in inhibition of growth and induction of cell death [22–30]. Exogenous expression of IL-24 using mesenchymal stem cells (MSCs) was shown to inhibit the growth of glioma and melanoma [31, 32]. While all of the above-described studies demonstrated the antitumor activity of IL-24, the molecular mechanism by which IL-24 induced tumor cell death varied and was shown to be cell type-dependent. Pataer et al. [33] reported IL-24-mediated cell death in lung cancer cells occurred by activation of protein kinase R (PKR). In human ovarian cancer cells, the Fas signaling pathway was shown to be involved in IL-24-mediated cell killing [34]. Activation of the JNK pathway was observed in Ad-mda7-treated glioma cells [22]. In human melanoma, an inverse correlation between IL-24 and inducible nitric oxide synthase (iNOS) was observed and treatment of melanoma cell lines with Ad-mda7 or IL-24 protein suppressed iNOS [35]. Regulation of the beta-catenin/Wnt signaling pathway was reported in Ad-mda7-treated pancreatic, breast, and lung cancer cells [36, 37]. Panneerselvam et al. [38] showed phosphorylation of IL-24 is a prerequisite for IL-24-mediated antitumor activity. More recently, suppression of the oncogenic GLI1-hedgehog signaling was

reported in IL-24 overexpressing lung cancer cells [39]. While the upstream signaling mechanism for IL-24-mediated killing differs among various cancer cell lines, the signals converge downstream at the mitochondria leading to activation of caspase-mediated apoptosis [40].

Apart from apoptotic-mediated cell death, involvement of autophagy in Ad-mda7-treated cells has also been reported [41–43]. In glioma, Ad-mda7 treatment activated protein kinase R-like endoplasmic reticulum kinase (PERK) leading to autophagy [44]. Further, combining Ad-mda7 with OSU-03012, an inducer of autophagy, increased ER stress and autophagy resulting in enhanced antitumor activity in gliomas [45]. Conversely, inhibition of autophagy using 3-methyl adenine (3-MA) enhanced IL-24-mediated cell death in human oral squamous cell carcinoma cell lines [46]. All of these studies show that both, autophagy and apoptotic cell death, are involved in IL-24 treatment.

7.2.1   Intracellular and Extracellular Protein-Mediated Cell Death

Since IL-24 protein has a secretory signal and is shown to be secreted into the extracellular environment, the role of its receptors in the mode of cell killing has been interrogated. Blocking IL-24 secretion did not abrogate tumor cell killing indicating intracellular protein expression is sufficient to exert the antitumor activity [47]. Intracellular accumulation of the IL-24 protein resulted in unfolded protein response (UPR) leading to endoplasmic reticulum (ER) stress culminating in cell death [47, 48]. In a separate study, involvement of both intra- and extracellular-mediated killing was reported in Ad-mda7-treated tumor cells [49]. Involvement of IL-20 receptor-mediated Ad-IL-24 cell killing was observed in melanoma and breast cancer cells [50, 51]. These studies showed that secreted IL-24 protein, upon binding to IL-20 receptor (IL-20R), activated signal transducer and transactivation (STAT-3) signaling. However, cell killing occurred independent of STAT-3 as inhibiting STAT-3 did not abrogate Ad-IL-24-mediated cell killing. In a separate study, addition of soluble IL-24 protein to human umbilical vein endothelial cells (HUVEC) and human microvascular endothelial cells (HMVEC) was shown to inhibit capillary tube formation [52]. Molecular studies showed IL-24 protein bound to its receptors to exert the inhibitory activity on capillary tube formation.

This demonstrates that extracellularly secreted IL-24 protein can exert its antitumor activity utilizing either of the two receptors (IL-20R and IL-22R) expressed by tumor and tumor-associated endothelial cells. The results from all of the studies described above showed that IL-24 potently exerts its antitumor activity by utilizing both intracellular and extracellular signaling mechanisms, in an autocrine and paracrine manner.

7.2.2 Combinatorial Therapy Enhances Tumor Cell Death

While early studies focused on testing IL-24 as monotherapy, it is imperative to test combinatorial therapies to reflect clinical relevance [53]. Combining Ad-IL-24 with geftinib and trastuzumab, inhibitors of epidermal growth factor receptor (EGFR) and Her2-neu receptor, respectively, demonstrated enhanced antitumor activity in lung and breast cancer models [54–56]. Similarly, combination therapy of Ad-mda7 with the anti-vascular endothelial growth factor (VEGF) antibody, bevacizumab, synergistically inhibited lung tumor growth both in vitro and in vivo [57]. Ad-mda7 when combined with the non-steroidal anti-inflammatory drug, sulindac, displayed synergistic antitumor activity against lung tumor cells [58]. Shanker et al. also demonstrated that vitamin E succinate upon combination with Ad-mda7 produced a pronounced anticancer activity against ovarian cancer cells [59]. Similarly, an additive to synergistic efficacy was produced when Ad-mda-7/IL-24 was combined with inhibitors against NFκB, Mcl1, and heat shock protein-90 (Hsp-90) [60–62]. Combined treatment of histone deacetylase (HDAC) inhibitors with Ad-mda7 produced

greater antitumor activity against pancreatic cancer cells and glioma [63, 64]. Combining conventional chemotherapeutics such as doxorubicin and temozolomide with Ad-mda7 also improved the antitumor efficacy against osteosarcoma and melanoma [65, 66]. Finally, radiation therapy in combination with Ad-mda7 not only increased DNA damage and greater antitumor activity but also inhibited angiogenesis and reverted drug-resistant tumor cells to temozolomide treatment [67, 68].

7.3 IL-24 as an Inhibitor of Tumor Angiogenesis

The initial report on the anti-angiogenic activity of IL-24 stemmed from the reduced CD31-positive endothelial vessel staining in Ad-mda7-treated lung tumor tissue and inhibition of endothelial cells to form capillary tubes in vitro [19]. This led to the testing of the antiangiogenic activity of Ad-mda7 in in vitro [52, 69] as well as in in vivo studies [52, 57, 67]. Ad-mda7 treatment of prostate tumor cells reduced vascular endothelial growth factor (VEGF) expression by inhibiting the Src kinase activity [69]. Concurring with these results was the report by Nishikawa et al. who showed Ad-mda7 reduced VEGF-mediated angiogenesis when combined with radiation therapy [67]. In a follow-up study, combining Ad-mda7 with the anti-VEGF inhibitor, bevacizumab, almost completely abolished VEGF in the extracellular environment and significantly suppressed tumor growth in vivo [57]. However, the key evidence for the anti-angiogenic activity of IL-24 was revealed using soluble IL-24 protein in in vitro and in vivo studies [52]. In vitro, addition of soluble IL-24 protein to HUVEC and HMVEC inhibited VEGF and basic fibroblast growth factor (bFGF)-induced endothelial cell differentiation to form capillary tubes, and cell migration in a dose-dependent manner. The antiangiogenic activity exerted by IL-24 was found to be more potent than known antiangiogenic agents such as endostatin, interferon gamma (IFN-γ), and IFN-inducible protein-10 (IP-10). That the inhibitory activity occurred via the IL-22 receptor was demonstrated using an anti-IL-22 antibody that abrogated the inhibitory activity of IL-24 on HUVEC capillary tube formation. In vivo, subcutaneous (s.c.) implantation of a co-mixture of A549 tumor cells (which express low levels of IL-24 receptors) with human embryonic kidney-293 (HEK293) stably transfected with human IL-24 cDNA and producing soluble IL-24, not only suppressed tumor growth established in the lower right flank of nude mice, but also suppressed the growth of the contralateral tumor established on the left lower flank of mice. *Note*, the contralateral tumor comprised of only A549 lung tumor cells. This result indicated that soluble IL-24 protein detected in the circulating blood was produced from the right flank tumor and inhibited the contralateral tumor by suppressing tumor angiogenesis. This was the first study that demonstrated the direct antiangiogenic activity for IL-24.

7.4 IL-24 as an Inhibitor of Cell Invasion and Metastasis

Establishment of tumor metastasis in a different location of the same organ (e.g., contralateral lobes of the lung) or at a distant site in a different organ (e.g., lung tumor metastasizing to the brain) requires tumor cell extravasation and intravasation and involves a series of finely orchestrated cellular and molecular processes. Key to these processes includes tumor cell migration and invasion. Studies showed Ad-mda7 treatment reduced focal adhesion kinase (FAK) activity resulting in significant inhibition of lung tumor cell migration and invasion in vitro [70]. In a follow-up study, Panneerselvam et al. [71] using a doxycycline-inducible system showed that induction of IL-24 protein inhibited cell migration of lung cancer cells by disrupting the chemokine receptor-4 (CXCR-4)/stromal-derived factor (SDF) axis. In the same study, the authors showed that incorporating CXCR4 inhibitors such as AMD3100 and SJA5 greatly enhanced the suppressive activity of IL-24 on tumor cell migration. Molecular studies showed IL-24 post-transcriptionally reduced the half-life

CXCR4 mRNA resulting in decreased protein expression. These important observations provide opportunities for testing new combinatorial therapies with IL-24-therapy for eliminating the metastatic properties of tumor cells thereby offering improved treatment outcomes and increased survival benefits.

7.5 IL-24 as an Inhibitor of Cancer Stem Cells

The presence of cancer stem cells (CSCs) or cancer stem-like cells, albeit low in number, within an actively growing tumor pose challenges in cancer treatment. CSCs are relatively resistant to conventional therapies and are hard to detect and eliminate resulting in their contribution in drug resistance, disease relapse, metastasis, and death. Therefore, availability of therapeutics that can effectively eliminate CSCs will reduce disease relapse and help to achieve better treatment outcomes. In this context, IL-24 has been tested for its cytotoxicity against CSCs. The first report on the ability of IL-24 to kill CSCs was demonstrated in a breast cancer model [72] in which proliferation of CSCs was suppressed by Ad-mda7, in vitro, by inhibiting the beta-catenin/Wnt signaling. Furthermore, in vivo studies showed growth of tumors generated from CSCs was effectively controlled upon Ad-mda7 treatment. The results from this study were subsequently validated in two additional, but separate set of studies, using prostate and laryngeal cancer models [73, 74]. In the prostate cancer model, IL-24 expression reduced the stemness of prostate cancer cells and sensitized the tumor cells to chemotherapy [73]. In the laryngeal tumor model, IL-24 reduced the CD133+ cells indicative of suppression of CSC or tumor initiating cell proliferation [74]. While additional validation of these study results is warranted, it is nevertheless evident that IL-24 does exert an inhibitory effect on CSCs, an important and hard to eradicate cell population present within the tumor microenvironment.

7.6   IL-24 as an Activator of Immune Cell Function

The first evidence for IL-24 possessing cytokine function was reported in 2002 by Caudell et al. [75]. The authors could detect IL-24 protein expression in human peripheral blood mononuclear cells (PBMC) primarily in CD19+ B cells and CD56+ natural killer (NK) cells. Further, treatment of PBMC with lipopolysaccharide (LPS) or phytohemagglutinin (PHA) resulted in increased IL-24 protein expression in the CD9+ and CD56+ subset of lymphocytes. Addition of soluble IL-24 protein to PBMCs resulted in the production and secretion of proinflammatory cytokines such as tumor necrosis factor-alpha (TNFα), interleukin 6 (IL-6), and interferon gamma (IFN-γ). Additionally, secretion of IL-12 and GMCSF was also noted albeit less than TNFα, IL-6, and IFN-γ. The cytokine-activating effect of IL-24 on PBMCs, however, was completely abrogated in the presence of IL-10 protein thereby resulting in a loss of TNFα, IFN-γ, IL-12, and GMCSF. Loss of IL-6 and IL-1β, however, was observed to be partial. These data indicated that although both IL-24 and IL-10 belong to the same cytokine family, they have opposing effects and operate as type 1 (Th1) and type 2 (Th2) cytokines respectively. This study, while establishing the cytokine function of IL-24, did not conduct any experiments to demonstrate the ability of IL-24 to elicit immune cell-mediated tumor cell cytotoxicity.

The real test of IL-24's ability to function and elicit T-cell-mediated tumor cell cytotoxicity was conducted by Miyahara et al. [76]. In their study, treatment of murine UV2237m fibrosarcoma established in an immunocompetent syngeneic mouse model with Ad-mda7 not only suppressed tumor growth but completely eradicated tumors in few treated mice. Challenging these tumor-free mice with a second dose of tumor cells completely rejected tumor growth indicating the involvement of antitumor immunity. Furthermore, splenocytes isolated from vaccinated mice and tumor-free mice had higher proliferative rate and increased Th1 cytokine (TNFα, IL-1, IFN-γ, IL-12) expression compared to splenocytes from non-vaccinated mice or those that had progressive tumor growth. Finally, phenotypic characterization of T-cell subsets from splenocytes of Ad-mda7-treated mice revealed an increase in the number of CD3+/CD8+ T cells indicating activation of cytotoxic CD8+ T cells, important for eradicating actively growing tumor. This study clearly established the ability of IL-24 to function as a Th1 cytokine and activate memory T cells required for effectively eradicating residual disease. Finally, this study showed that IL-24 functions as an anticancer vaccine thereby paving the way for further characterization and testing of IL-24 as a cancer therapeutic.

In two independent follow-up studies conducted using colon cancer model, IL-24 was reported to function as a Th1 cytokine and activate the host immune response resulting in suppression of tumor growth [77, 78]. Ma et al. [77] showed recombinant GST-IL-24 fusion protein treatment altered the tumor microenvironment into a robust immune-activated environment that resulted in heightened number of CD4+ and CD8+ T-cell population in CT-24-colon tumor-bearing mice. Additionally, increased IFN production was observed in CD8+ T cells isolated from the splenocytes of GST-IL-24-treated tumor-bearing mice. Analysis of tumor infiltrating lymphocytes (TILs) showed GST-IL-24 reduced FoxP3+ T-regulatory cells (TRegs) while concomitantly increasing CD45+/CD4+ and CD45+/CD8+ T cells. In vitro cytotoxicity T-cell assay showed CD8+ T cells but not CD4+ cells effectively killed tumor cells. Intravenous (i.v.) administration of GST-IL24 protein to CT26 colon tumor-bearing mice showed slowing of tumor growth compared to untreated control mice. Depletion of CD8+ and CD4+ cells from the tumor-bearing mice prior to GST-IL-24 treatment showed loss of tumor-suppressive activity in CD8+ depleted mice but not in CD4+ depleted mice clearly indicating that CD8+ T cells are required for IL-24-mediated immune activity. Finally, colon cancer patients showed a strong correlation between IL-24 expression, CD4+/CD8+ T-cell population, and tumor burden. Loss or reduced IL-24 expression correlated with a higher number of Fox3P+ T cells and higher grade of colon cancer. Inversely, high IL-24 expression correlated with higher CD4+/CD8+ population and lower tumor burden and better survival outcomes.

Similar observation was made by Zhang et al. [70], who compared blood and tissue samples from 29 patients diagnosed with colon cancer and 15 normal healthy individuals. mRNA expression for IL-24 was decreased in colon cancer tissues. In contrast, no difference in IL-24 receptor expression was observed between colon cancer patients and healthy individuals. TILs isolated from the tumors and treated with low concentration of IL-24 protein suppressed CD4+ proliferation and reduced cytokine production. In contrast, treatment of TILs with higher concentration of IL-24 protein activated CD4+ cell proliferation and increased cytokine production along with a concomitant decrease in Fox3P+ TReg cell population. Finally, high IL-24 protein concentration boosted the cytotoxic activity of CD8+ T cells while lower IL-24 protein concentration had no effect demonstrating the need for a critical amount of IL-24 to be present within the tumor microenvironment to elicit a robust antitumor immune response.

Further, a phase I clinical study was conducted to test Ad-mda7 for the treatment of solid tumors. Intratumoral (i.t.) injection of Ad.5-mda-7 (INGN-241) was found to be safe and exhibited significant clinical efficacy in approximately 44% of the cancer patients treated with multiple doses of Ad-mda7 [79, 80]. Analysis of pre- and post-treatment tumor tissue samples post Ad-mda7 treatment showed IL-24 protein expression and induction of tumor cell apoptosis as measured by TUNEL staining. Furthermore, change in cytokine profile and CD4+/CD8+ cell population was observed following Ad-mda7 treatment. The results from this clinical study convincingly demonstrated Ad-mda7 operated both, as a tumor suppressor and a Th1 cytokine, inducing apoptotic cell death and stimulating a strong antitumor immune response [81].

In summary, IL-24 has multifunctional (antitumor, antiangiogenic, antimetastatic, and proimmune) properties (Fig. 7.1) that warrant its development as an anticancer drug for treating cancer patients.

7.7 Concluding Remarks

From the time of its discovery and the first report about its antitumor activity in early 1990s, there has been tremendous interest to investigate and develop IL-24 as a bio-therapeutic for cancer treatment. While results from preclinical studies have led to the testing of the antitumor activity of IL-24 in a phase I clinical trial, gaps in knowledge about its mode of action exist. For example, how endogenous IL-24 expression is regulated in normal and tumor cells remains to be explored. Though the antitumor activity of IL-24 is well-studied, its function as a tumor suppressor and a cytokine and the signaling mechanisms it employs to separate the two biological functions are unknown. While the IL-24-mediated antiangiogenic activity has been documented using endothelial cells as a model, the impact of IL-24 on pericytes especially of the type 2 phenotype is unknown [82]. Similarly, testing of IL-24 in the context of immune-based studies is limited. Finally, testing and advancing IL-24 protein-based cancer therapy will be beneficial for clinical application. Addressing these questions will unfold new and exciting information that will enable scientists to successfully reconfigure the tumor microenvironment to elicit a strong antitumor immune response and improve IL-24-based cancer therapy.

Acknowledgments This study was supported by grants received from National Institutes of Health (NIH) (R01 CA167516 and R01CA233201); a Merit Grant (101BX003420A1) from the Department of Veterans Affairs; a Seed Grant and Trainee Support Grant funded by the National Cancer Institute Cancer Center Support Grant P30CA225520 awarded to the University of Oklahoma Stephenson Cancer Center; a grant (HR18-088) from the Oklahoma Center for Advanced Science and Technology (OCAST); funds received from the Presbyterian Health Foundation (PHF) Seed Grant, Presbyterian Health Foundation Bridge Grant, and the Jim and Christy Everest Endowed Chair in Cancer Developmental Therapeutics.

The content presented is solely the responsibility of the authors. The opinions, interpretations, conclusions,

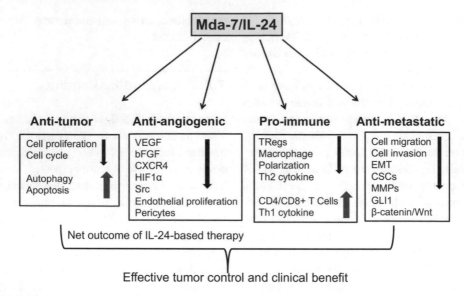

Fig. 7.1 Schematic showing the multifunctional properties of mda-7/IL-24 inhibiting several components essential for cancer growth and metastasis

and recommendations are those of the author and not necessarily endorsed by or representative of the official views of NIH, Department of Veterans Affairs, OCAST, or PHF.

Rajagopal Ramesh is an Oklahoma TSET Research Scholar and holds the Jim and Christy Everest Endowed Chair in Cancer Developmental Therapeutics.

Conflict of interest: The authors report no conflict of interest in this work.

References

1. Inoue S, Shanker M, Miyahara R, Gopalan B, Patel S, Oida Y, Branch CD, Munshi A, Meyn RE, Andreeff M, Tanaka F, Mhashilkar AM, Chada S, Ramesh R (2006) MDA-7/IL-24-based cancer gene therapy: translation from the laboratory to the clinic. Curr Gene Ther 6:73–91

2. Fisher PB, Gopalkrishnan R, Chada S, Ramesh R, Grimm EΛ, Rosenfeld MR, Curiel DT, Dent P (2003) *mda*-7/IL-24, novel cancer selective apoptosis inducing cytokine gene. From the laboratory into the clinic. Cancer Biol Ther 2:S23–S37

3. Huang EY, Madireddi MT, Gopalkrishnan RV, Leszczyniecka M, Su Z, Lebedeva IV, Kang D, Jiang H, Lin JJ, Alexandre D, Chen Y, Vozhilla N, Mei MX, Christiansen KA, Sivo F, Goldstein NI, Mhashilkar AB, Chada S, Huberman E, Pestka S, Fisher PB (2001) Genomic structure, chromosomal localization and expression profile of a novel melanoma differentiation associated (mda-7) gene with cancer specific growth suppressing and apoptosis inducing properties. Oncogene 20:7051–7063

4. Niess JH, Francés R (2019) Editorial: the IL-20 cytokines and related family members in immunity and diseases. Front Immunol 10:1976. https://doi.org/10.3389/fimmu.2019.01976

5. Ouyang W, O'Garra A (2019) IL-10 family cytokines IL-10 and IL-22: from basic science to clinical translation. Immunity 50:871–891. https://doi.org/10.1016/j.immuni.2019.03.020

6. Gallagher G, Dickensheets H, Eskdale J, Izotova LS, Mirochnitchenko OV, Peat JD et al (2000) Cloning, expression and initial characterization of interleukin-19 (IL-19), a novel homologue of human interleukin-10 (IL-10). Genes Immun 1:442–450. https://doi.org/10.1038/sj.gene.6363714

7. Blumberg H, Conklin D, Xu WF, Grossmann A, Brender T, Carollo S et al (2001) Interleukin 20: discovery, receptor identification, and role in epidermal function. Cell 104:9–19

8. Jiang H, Lin JJ, Su ZZ, Goldstein NI, Fisher PB (1995) Subtraction hybridization identifies a novel melanoma differentiation associated gene, mda-7, modulated during human melanoma differentiation, growth and progression. Oncogene 11:2477–2486

9. Mhashilkar AM, Schrock RD, Hindi M, Liao J, Sieger K, Kourouma F, Zou-Yang XH, Onishi E, Takh O, Vedvick TS, Fanger G, Stewart L, Watson GJ, Snary D, Fisher PB, Saeki T et al (2001) Melanoma differentiation associated gene-7 (mda-7): a novel anti-tumor gene for cancer gene therapy. Mol Med 7:271–282

10. Fuson KL, Zheng M, Craxton M, Pataer A, Ramesh R, Chada S, Sutton RB (2009) Structural mapping of post-translational modifications in human interleukin-24: role of N-linked glycosylation and disulfide bonds in secretion and activity. J Biol Chem 284:30526–30533

11. Gupta P, Su ZZ, Lebedeva IV, Sarkar D, Sauane M, Emdad L, Bachelor MA, Grant S, Curiel DT, Dent P, Fisher PB (2006) mda-7/IL-24: multifunctional cancer-specific apoptosis-inducing cytokine. Pharmacol Ther 111:596–628

12. Chada S, Bocangel D, Pataer A, Sieger K, Mhashilkar AM, Inoue S, Miyahara R, Roth JA, Swisher S, Hunt KK, Ramesh R (2007) MDA-7/IL-24 as a multimodality therapy for cancer. In: Hunt KK, Vorburger S, Swisher SG (eds) Cancer drug discovery and development: gene therapy for cancer. Humana Press, USA, pp 413–433

13. Chada S, Sutton RB, Ekmekcioglu S, Ellerhorst J, Mumm JB, Leitner WW, Yang HY, Sahin AA, Hunt KK, Fuson KL, Poindexter N, Roth JA, Ramesh R, Grimm EA, Mhashilkar AM (2004) MDA-7/IL-24 is a unique cytokine—tumor suppressor in the IL-10 family. Int Immunopharmacol 4:649–667

14. Gopalan B, Shanker M, Scott A, Branch CD, Chada S, Ramesh R (2008) MDA-7/IL-24, a novel tumor suppressor/cytokine is ubiquitinated and regulated by the ubiquitin-proteasome system, and inhibition of MDA-7/IL-24 degradation enhances the antitumor activity. Cancer Gene Ther 15:1–8

15. Dumoutier L, Leemans C, Lejeune D, Kotenko SV, Renauld JC (2001) Cutting edge: STAT activation by IL-19, IL-20 and mda-7 through IL-20 receptor complexes of two types. J Immunol 167:3545–3549

16. Wang M, Tan Z, Zhang R, Kotenko SV, Liang P (2002) Interleukin 24 (MDA-7/MOB-5) signals through two heterodimeric receptors, IL-22R1/IL-20R2 and IL-20R1/IL-20R2. J Biol Chem 277:7341–7347

17. Su ZZ, Madireddi MT, Lin JJ, Young CS, Kitada S, Reed JC, Goldstein NI, Fisher PB (1998) The cancer growth suppressor gene mda-7 selectively induces apoptosis in human breast cancer cells and inhibits tumor growth in nude mice. Proc Natl Acad Sci U S A 95:14400–14405

18. Saeki T, Mhashilkar A, Chada S, Branch C, Roth JA, Ramesh R (2000) Tumor-suppressive effects by adenovirus-mediated mda-7 gene transfer in non-small cell lung cancer cell in vitro. Gene Ther 7:2051–2057

19. Saeki T, Mhashilkar A, Swanson X, Zou-Yang XH, Sieger K, Kawabe S, Branch CD, Zumstein L, Meyn RE, Roth JA, Chada S, Ramesh R (2002) Inhibition of human lung cancer growth following adenovirus-mediated mda-7 gene expression in vivo. Oncogene 21:4558–4566

20. Ellerhorst JA, Prieto VG, Ekmekcioglu S, Broemeling L, Yekell S, Chada S, Grimm EA (2002) Loss of MDA-7 expression with progression of melanoma. J Clin Oncol 20:1069–1074

21. Ishikawa S, Nakagawa T, Miyahara R, Kawano Y, Takenaka K, Yanagihara K, Otake Y, Katakura H, Wada H, Tanaka F (2005) Expression of MDA-7/IL-24 and its clinical significance in resected non-small cell lung cancer. Clin Cancer Res 11:1198–1202

22. Yacoub A, Mitchell C, Lebedeva IV, Sarkar D, Su ZZ, McKinstry R, Gopalkrishnan RV, Grant S, Fisher PB, Dent P (2003) mda-7 (IL-24) inhibits growth and enhances radiosensitivity of glioma cells in vitro via JNK signaling. Cancer Biol Ther 2:347–353

23. Ramesh R, Ito I, Miyahara R, Saito Y, Wu Z, Mhashilkar AM, Wilson DR, Branch CD, Chada S, Roth JA (2004) Local and systemic inhibition of lung tumor growth after nanoparticle-mediated mda-7/IL-24 gene delivery. DNA Cell Biol 23:850–857

24. Liu L, Ma J, Qin L, Shi X, Si H, Wei Y (2019) Interleukin-24 enhancing antitumor activity of chimeric oncolytic adenovirus for treating acute promyelocytic leukemia cell. Medicine (Baltimore) 98:e15875

25. Luo H, Chen Y, Niu L, Liang A, Yang J, Li M (2019) Hepatoma cell-targeted cationized silk fibroin as a carrier for the inhibitor of growth 4-interleukin-24 double gene plasmid. J Biomed Nanotechnol 15:1622–1635

26. Gopalan B, Shanker M, Chada S, Ramesh R (2007) MDA-7/IL-24 suppresses human ovarian carcinoma growth in vitro and in vivo. Mol Cancer 6:11

27. Deng L, Fan J, Ding Y, Yang X, Huang B, Hu Z (2020) Target therapy with vaccinia virus harboring IL-24 for human breast cancer. J Cancer 11:1017–1026

28. Xu X, Yi C, Yang X, Xu J, Sun Q, Liu Y, Zhao L (2019) Tumor cells modified with Newcastle disease virus expressing IL-24 as a cancer vaccine. Mol Ther Oncolytics 14:213–221

29. Mo Q, Liu L, Bao G, Li T (2019) Effects of melanoma differentiation associated gene-7 (MDA-7/IL-24) on apoptosis of liver cancer cells via regulating the expression of B-cell lymphoma-2. Oncol Lett 18:29–34

30. Maehana S, Matsumoto Y, Kojima F, Kitasato H (2019) Interleukin-24 transduction modulates human prostate cancer malignancy mediated by regulation of anchorage dependence. Anticancer Res 39:3719–3725

31. Wu Z, Liu W, Wang Z, Zeng B, Peng G, Niu H, Chen L, Liu C, Hu Q, Zhang Y, Pan M, Wu L, Liu M, Liu X, Liang D (2020) Mesenchymal stem cells derived from iPSCs expressing interleukin-24 inhibit the growth of melanoma in the tumor-bearing mouse model. Cancer Cell Int 20:33. https://doi.org/10.1186/s12935-020-1112-7

32. Fan S, Gao H, Ji W, Zhu F, Sun L, Liu Y, Zhang S, Xu Y, Yan Y, Gao Y (2020) Umbilical cord-derived mesenchymal stromal/stem cells expressing IL-24 induce apoptosis in gliomas. J Cell Physiol 235:1769–1779

33. Pataer A, Vorburger SA, Chada S, Balachandran S, Barber GN, Roth JA, Hunt KK, Swisher SG (2005) Melanoma differentiation-associated gene-7 protein physically associates with the double-stranded RNA-activated protein kinase PKR. Mol Ther 11:717–723

34. Gopalan B, Litvak A, Sharma S, Mhashilkar AM, Chada S, Ramesh R (2005) Activation of the Fas-FasL signaling pathway by MDA-7/IL-24 kills human ovarian cancer cells. Cancer Res 65:3017–3024

35. Ekmekcioglu S, Ellerhorst JA, Mumm JB, Zheng M, Broemeling L, Prieto VG, Stewart AL, Mhashilkar AM, Chada S, Grimm EA (2003) Negative association of melanoma differentiation-associated gene (mda-7) and inducible nitric oxide synthase (iNOS)

in human melanoma: MDA-7 regulates iNOS expression in melanoma cells. Mol Cancer Ther 2:9–17

36. Chada S, Bocangel D, Ramesh R, Grimm EA, Munshi A, Meyn RE, Mhashilkar AM, Zheng M (2005) mda7 kills pancreatic cancer cells by inhibition of the Wnt/PI3K signaling pathways: identification of IL20 receptor-mediated bystander activity against pancreatic cancer. Mol Ther 11:724–733

37. Mhashilkar AM, Stewart AL, Sieger K, Yang HY, Khimani AH, Ito I, Saito Y, Hunt KK, Grimm EA, Roth JA, Meyn RE, Ramesh R, Chada S (2003) MDA-7 negatively regulates the beta-catenin and PI3K signaling pathways in breast and lung tumor cells. Mol Ther 8:207–219

38. Panneerselvam J, Shanker M, Jin J, Branch CD, Muralidharan R, Zhao YD, Chada S, Munshi A, Ramesh R (2015) Phosphorylation of interleukin (IL)-24 is required for mediating its anti-cancer activity. Oncotarget 6:16271–16286

39. Panneerselvam J, Srivastava A, Mehta M, Chen A, Zhao YD, Munshi A, Ramesh R (2019) IL-24 inhibits lung cancer growth by suppressing GLI1 and inducing DNA damage. Cancers (Basel) 11(12):E1879. https://doi.org/10.3390/cancers11121879

40. Panneerselvam J, Munshi A, Ramesh R (2013) Molecular targets and signaling pathways regulated by interleukin (IL)-24 in mediating its antitumor activities. J Mol Signal 8:15

41. Yokoyama T, Miyamoto S, Ramesh R (2010) Interleukin (IL)-24: a regulator of autophagy and apoptosis-mediated programmed cell death. Trends Cell Mol Biol 5:61–67

42. Lehman S, Koumenis C (2010) The role of autophagy as a mechanism of cytotoxicity by the clinically used agent MDA-7/IL-24. Cancer Biol Ther 9:537–538

43. Emdad L, Bhoopathi P, Talukdar S, Pradhan AK, Sarkar D, Wang XY, Das SK, Fisher PB (2019) Recent insights into apoptosis and toxic autophagy: the roles of MDA-7/IL-24, a multidimensional anticancer therapeutic. Semin Cancer Biol 66:140–154

44. Park MA, Yacoub A, Sarkar D, Emdad L, Rahmani M, Spiegel S, Koumenis C, Graf M, Curiel DT, Grant S, Fisher PB, Dent P (2008) PERK-dependent regulation of MDA-7/IL-24-induced autophagy in primary human glioma cells. Autophagy 4:513–515

45. Hamed HA, Yacoub A, Park MA, Eulitt P, Sarkar D, Dimitrie IP, Chen CS, Grant S, Curiel DT, Fisher PB, Dent P (2010) OSU-03012 enhances Ad.7-induced GBM cell killing via ER stress and autophagy and by decreasing expression of mitochondrial protective proteins. Cancer Biol Ther 9:526–536

46. Li J, Yang D, Wang W, Piao S, Zhou J, Saiyin W, Zheng C, Sun H, Li Y (2015) Inhibition of autophagy by 3-MA enhances IL-24-induced apoptosis in human oral squamous cell carcinoma cells. J Exp Clin Cancer Res 34:97

47. Sieger KA, Mhashilkar AM, Stewart A, Sutton RB, Strube RW, Chen SY, Pataer A, Swisher SG, Grimm EA, Ramesh R, Chada S (2004) The tumor suppressor activity of MDA-7/IL-24 is mediated by intracel-

lular protein expression in NSCLC cells. Mol Ther 9:355–367

48. Peñaranda Fajardo NM, Meijer C, Kruyt FA (2016) The endoplasmic reticulum stress/unfolded protein response in gliomagenesis, tumor progression and as a therapeutic target in glioblastoma. Biochem Pharmacol 118:1–8

49. Sauane M, Lebedeva IV, Su ZZ, Choo HT, Randolph A, Valerie K, Dent P, Gopalkrishnan RV, Fisher PB (2004) Melanoma differentiation associated gene-7/interleukin-24 promotes tumor cell-specific apoptosis through both secretory and nonsecretory pathways. Cancer Res 64:2988–2993

50. Chada S, Mhashilkar AM, Ramesh R, Mumm JB, Sutton RB, Bocangel D, Zheng M, Grimm EA, Ekmekcioglu S (2004) Bystander activity of Ad-mda7: MDA-7 protein kills melanoma cells via an IL-20 receptor-dependent but STAT-3-independent mechanism. Mol Ther 10:1085–1095

51. Zheng M, Bocangel D, Doneske B, Mhashilkar A, Ramesh R, Hunt KK, Ekmekcioglu S, Poindexter N, Grimm EA, Chada S (2007) Human interleukin 24 (IL-24) protein kills breast cancer cells via the IL-20 receptor and is antagonized by IL-10. Cancer Immunol Immunother 56:205–215

52. Ramesh R, Mhashilkar AM, Tanaka F, Saito Y, Branch CD, Sieger K, Mumm JB, Stewart AL, Boquoi A, Dumoutier L, Grimm EA, Renauld JC, Kotenko S, Chada S (2003) Melanoma differentiation-associated gene 7/interleukin (IL)-24 is a novel ligand that regulates angiogenesis via the IL-22 receptor. Cancer Res 63:5105–5113

53. Chada S, Mhashilkar AM, Liu Y, Nishikawa T, Bocangel D, Zheng M, Vorburger SA, Pataer A, Swisher SG, Ramesh R, Kawase K, Meyn RE, Hunt KK (2006) Mda-7 gene transfer sensitizes breast carcinoma cells to chemotherapy, biologic therapies and radiotherapy: correlation with expression of bcl-2 family members. Cancer Gene Ther 13:490–502

54. Emdad L, Lebedeva IV, Su ZZ, Gupta P, Sarkar D, Settleman J, Fisher PB (2007) Combinatorial treatment of non-small-cell lung cancers with gefitinib and Ad.mda-7 enhances apoptosis-induction and reverses resistance to a single therapy. J Cell Physiol 210:549–559

55. McKenzie T, Liu Y, Fanale M, Swisher SG, Chada S, Hunt KK (2004) Combination therapy of Ad-mda7 and trastuzumab increases cell death in Her-2/neu-overexpressing breast cancer cells. Surgery 136:437–442

56. Bocangel D, Zheng M, Mhashilkar A, Liu Y, Hunt KK, Ramesh R, Chada S (2006) Combinatorial synergy induced by Ad-mda7 and Herceptin in Her2+ breast cancer cells. Cancer Gene Ther 13:958–968

57. Inoue S, Hartman A, Branch CD, Bucana CD, Bekele NB, Stephens LC, Chada S, Ramesh R (2007) MDA-7 in combination with Bevacizumab treatment produces a synergistic and complete inhibitory effect on lung tumor xenograft. Mol Ther 15:287–294

58. Oida Y, Gopalan B, Miyahara R, Inoue S, Branch C, Mhashilkar AM, Roth JA, Chada S, Ramesh R (2005) Sulindac enhances Ad-mda7/IL-24 mediated apoptosis of human lung cancer cells. Mol. Cancer Ther 4:291–304

59. Shanker M, Gopalan B, Patel S, Bocangel D, Chada S, Ramesh R (2007) Vitamin E succinate in combination with ad-mda7 modulates the extrinsic and intrinsic apoptotic pathways resulting in enhanced human ovarian tumor cell killing. Cancer Lett 254:217–226

60. Oida Y, Gopalan B, Miyahara R, Branch CD, Chiao P, Chada S, Ramesh R (2007) Inhibition of nuclear factor kappa B (NFkB) augments antitumor activity of Adenovirus-mediated melanoma differentiation associated gene-7 against lung cancer cells via MEKK1. Mol Cancer Ther 6:1440–1449

61. Dash R, Azab B, Quinn BA, Shen X, Wang XY, Das SK, Rahmani M, Wei J, Hedvat M, Dent P, Dmitriev IP, Curiel DT, Grant S, Wu B, Stebbins JL, Pellecchia M, Reed JC, Sarkar D, Fisher PB (2011) Apogossypol derivative BI-97C1 (Sabutoclax) targeting Mcl-1 sensitizes prostate cancer cells to mda-7/IL-24-mediated toxicity. Proc Natl Acad Sci U S A 108:8785–8790

62. Zhang Z, Kawamura K, Jiang Y, Shingyoji M, Ma G, Li Q, Hu J, Qi Y, Liu H, Zhang F, Kang S, Shan B, Wang S, Chada S, Tagawa M (2013) Heat -shock protein 90 inhibitors synergistically enhance melanoma differentiation-associated gene-7-mediated cell killing of human pancreatic carcinoma. Cancer Gene Ther 20:663–670

63. Hamed HA, Yacoub A, Park MA, Archer K, Das SK, Sarkar D, Grant S, Fisher PB, Dent P (2013) Histone deacetylase inhibitors interact with melanoma differentiation associated-7/interleukin-24 to kill primary human glioblastoma cells. Mol Pharmacol 84:171–181

64. Hamed HA, Das SK, Sokhi UK, Park MA, Cruickshanks N, Archer K, Ogretmen B, Grant S, Sarkar D, Fisher PB, Dent P (2013) Combining histone deacetylase inhibitors with MDA-7/IL-24 enhances killing of renal carcinoma cells. Cancer Biol Ther 14:1039–1049

65. Liu Z, Xu L, Yuan H, Zhang Y, Zhang X, Zhao D (2015) Oncolytic adenovirus-mediated mda-7/IL-24 expression suppresses osteosarcoma growth and enhances sensitivity to doxorubicin. Mol Med Rep 12:6358–6364

66. Zheng M, Bocangel D, Ramesh R, Ekmekcioglu S, Poindexter N, Grimm EA, Chada S (2008) IL-24 overcomes TMZ-resistance and enhances cell death by indirect downregulation of MGMT in human melanoma cells. Mol Cancer Ther 7:3842–3851

67. Nishikawa T, Ramesh R, Chada S, Meyn RE (2004) Suppression of tumor growth and angiogenesis by adenovirus-mediated mda-7/IL-24 gene transfer in combination with ionizing radiation. Mol Ther 9:818–828

68. Germano IM, Emdad L, Qadeer ZA, Binello E, Uzzaman M (2010) Embryonic stem cell (ESC)-mediated transgene delivery induces growth suppression, apoptosis and radiosensitization, and overcomes temozolomide resistance in malignant gliomas. Cancer Gene Ther 17:664–674

69. Inoue S, Branch CD, Gallick G, Chada S, Ramesh R (2005) Inhibition of Src kinase activity by Ad-mda7 suppresses vascular endothelial growth factor expression in prostate carcinoma cells. Mol Ther 12:707–715

70. Ramesh R, Ito I, Gopalan B, Saito Y, Mhashilkar AM, Chada S (2004) Ectopic production of MDA-7/IL-24 inhibits invasion and migration of human lung cancer cells. Mol Ther 9:510–518

71. Panneerselvam J, Jin J, Shanker M, Lauderdale J, Bates J, Wang Q, Zhao YD, Archibald SJ, Hubin TJ, Ramesh R (2015) IL-24 inhibits lung cancer cell migration and invasion by disrupting the SDF-1/CXCR4 signaling axis. PLoS One 10(3):e0122439

72. Bhutia SK, Das SK, Azab B, Menezes ME, Dent P, Wang XY, Sarkar D, Fisher PB (2013) Targeting breast cancer-initiating/stem cells with melanoma differentiation-associated gene-7/interleukin-24. Int J Cancer 133:2726–2736

73. Cheng JZ, Yu D, Zhang H, Jin CS, Liu Y, Zhao X, Qi XM, Liu XB (2014) Inhibitive effect of IL-24 gene on CD133(+) laryngeal cancer cells. Asian Pac J Trop Med 7:867–872

74. Yu D, Zhong Y, Li X, Li Y, Li X, Cao J, Fan H, Yuan Y, Ji Z, Qiao B, Wen JG, Zhang M, Kvalheim G, Nesland JM, Suo Z (2015) ILs-3, 6 and 11 increase, but ILs-10 and 24 decrease stemness of human prostate cancer cells in vitro. Oncotarget 6:42687–42703

75. Caudell EG et al (2002) The protein product of the tumor suppressor gene, melanoma differentiation-associated gene 7, exhibits immunostimulatory activity and is designated IL-24. J Immunol 168:6041–6046

76. Miyahara R, Banerjee S, Kawano K, Efferson C, Tsuda N, Miyahara Y, Ioannides CG, Chada S, Ramesh R (2006) Melanoma differentiation-associated gene-7 (mda-7)/interleukin (IL)-24 induces anticancer immunity in a syngeneic murine model. Cancer Gene Ther 13:753–761

77. Ma YF, Ren Y, Wu CJ, Zhao XH, Xu H, Wu DZ, Xu J, Zhang XL, Ji Y (2016) Interleukin (IL)-24 transforms the tumor microenvironment and induces anticancer immunity in a murine model of colon cancer. Mol Immunol 75:11–20

78. Zhang Y, Liu Y, Xu Y (2019) Interleukin-24 regulates T cell activity in patients with colorectal adenocarcinoma. Front Oncol 9:1401

79. Tong AW, Nemunaitis J, Su D, Zhang Y, Cunningham C, Senzer N, Netto G, Rich D, Mhashilkar A, Parker K, Coffee K, Ramesh R, Ekmekcioglu S, Grimm EA, van Wart HJ, Merritt J, Chada S (2005) Intratumoral injection of INGN 241, a nonreplicating adenovector expressing the melanoma-differentiation associated gene-7 (mda-7/IL24): biologic outcome in advanced cancer patients. Mol Ther 11:160–172

80. Cunningham CC, Chada S, Merritt JA, Tong A, Senzer N, Zhang Y, Mhashilkar A, Parker K, Vukelja S, Richards D, Hood J, Coffee K, Nemunaitis J (2005) Clinical and local biological effects of an intratumoral

injection of mda-7 (IL24; INGN 241) in patients with advanced carcinoma: a phase I study. Mol Ther 11:149–159

81. Lebedeva IV, Emdad L, Su ZZ, Gupta P, Sauane M, Sarkar D, Ramesh R, Inoue S, Chada S, Li R, DePass AL, Mahasreshti PJ, Curiel DT, Yacoub A, Grant S, Dent P, Nemunaitis JJ, Fisher PB (2008) mda-7/IL-24, novel anticancer cytokine: focus on bystander antitumor, radiosensitization and antiangiogenic properties and overview of the phase I clinical experience. Int J Oncol 31:985–1007

82. Birbrair A, Zhang T, Wang ZM, Messi ML, Olson JD, Mintz A, Delbono O (2014) Type-2 pericytes participate in normal and tumoral angiogenesis. Am J Physiol Cell Physiol 307:C25–C38

Interleukin-31, a Potent Pruritus-Inducing Cytokine and Its Role in Inflammatory Disease and in the Tumor Microenvironment

8

Alain H. Rook, Kathryn A. Rook, and Daniel J. Lewis

Abstract

Substantial new information has emerged supporting the fundamental role of the cytokine interleukin-31 (IL-31) in the genesis of chronic pruritus in a broad array of clinical conditions. These include inflammatory conditions, such as atopic dermatitis and chronic urticaria, to autoimmune conditions such as dermatomyositis and bullous pemphigoid, to the lymphoproliferative disorders of Hodgkin's disease and cutaneous T-cell lymphoma. IL-31 is produced in greatest quantity by T-helper type 2 (Th2) cells and upon release, interacts with a cascade of other cytokines and chemokines to lead to pruritus and to a proinflammatory environment, particularly within the skin. Antibodies which neutralize IL-31 or which block the IL-31 receptor may reduce or eliminate pruritus and may diminish the manifestations of chronic cutaneous conditions associated with elevated IL-31. The role of IL-31 in these various conditions will be reviewed.

Keywords

IL-31: Interleukin-31 · IL-33: Interleukin-33 · IL-4: Interleukin-4 · IL-13: Interleukin-13 · ILC2 cells: Type 2 innate immune cells · Atopic dermatitis · Dermatomyositis · Oclacitinib · Nemolizumab · Lokivetmab · Dupilumab · Interleukin-31 receptor · Chemokine · Th2: Type 2 helper T cell · Sézary syndrome · Mycosis fungoides · Cutaneous T-cell lymphoma (CTCL)

A. H. Rook (✉) · D. J. Lewis
Department of Dermatology, Perelman School of Medicine, University of Pennsylvania, Philadelphia, PA, USA
e-mail: arook@pennmedicine.upenn.edu; Daniel.Lewis@pennmedicine.upenn.edu

K. A. Rook
Clinical Dermatology and Allergy, University of Pennsylvania School of Veterinary Medicine, Philadelphia, PA, USA
e-mail: karook@vet.upenn.edu

Since its discovery more than 15 years ago, interleukin-31 (IL-31) has been implicated in the pathogenesis of a variety of immune-mediated conditions. Recent results suggest significant interactions of IL-31 with other type 2 (Th2) cytokines and chemokines in processes as diverse as cutaneous T-cell lymphoma (CTCL), follicular B-cell lymphoma, dermatomyositis, and atopic dermatitis. Importantly, the prominent role of IL-31 in causing pruritus has produced efforts to

A. Birbrair (ed.), *Tumor Microenvironment*, Advances in Experimental Medicine and Biology 1290, https://doi.org/10.1007/978-3-030-55617-4_8

target IL-31 directly as a strategy to ameliorate chronic pruritus. These topics will be briefly summarized in this chapter.

8.1   Introduction

IL-31 is a short-chain four-helix bundle cytokine that is produced by a variety of immune cell types, particularly by skin-homing CD4+ T cells. The initial discovery by Dillon and colleagues [1], through a functional cloning approach in cells co-expressing the heterodimeric IL-31 receptors, IL31RA and OSMRβ, supported observations that IL-31 was an important mediator in inflammatory skin disease, and, in particular, in the etiology of pruritus (Fig. 8.1). When overexpressed in transgenic mice, it was found to cause dermatitis, alopecia, and pruritus [1], with nearly incessant scratching behavior among the mice. These observations have served to stimulate significant investigation of the role of IL-31 in the pathogenesis of pruritus in numerous disease states which are inflammatory in nature as well as in malignancies associated with generalized pruritus.

8.2   Cellular Production of IL-31

Various cell types, particularly those of the immune system, have been observed to produce IL-31. Early work suggested that IL-31 was predominantly produced by CD4+/CD45RO+ T cells [2]. More recently, CD4+ skin-trafficking T cells have been found to be important sources of IL-31, which is highly relevant to cutaneous inflammatory conditions as well as to cutaneous lymphoproliferative processes mediated by malignant CD4+ T cells [3].

Many of the clinical conditions in which IL-31 has been found to play a role in the pathogenesis of pruritus and inflammation have been considered primarily Th2-driven immunologic processes. In this regard, IL-31 appears to be produced predominantly by Th2 T cells [1, 2]. However, Th1 T-cell clones can be induced in vitro to produce IL-31 in the presence of IL-4

demonstrating the important influence that the latter Th2 cytokine exerts on IL-31 production [4]. Included among the disorders with a Th2 bias is atopic dermatitis, which is considered a classic Th2-driven disease in which IL-4 and IL-13 are important mediators of the pathologic responses observed in the skin. Recent studies have suggested that IL-4 is a key stimulator of IL-31 production [4]. As discussed below, IL-31 is considered to play a critical role in the intense pruritus typically observed in severe atopic dermatitis [5]. Pruritic atopic patients treated with an IL-31 receptor-blocking antibody experience a significant reduction in pruritus, leading to reduced scratching behavior and improved disease manifestations [6]. These observations strongly support the importance of IL-31 in the symptoms and clinical manifestations of atopic dermatitis.

In addition, leukemic forms of CTCL, considered to be predominantly a malignancy of skin-trafficking CD4+/CLA+/CCR4+ T cells manifesting a Th2 bias, are also associated with elevated IL-31 protein in serum and IL-31 mRNA in skin lesions of patients with pruritus [7]. Improvement in pruritus with treatment of CTCL has been associated with reduced IL-31 levels at various disease sites, further indicating the importance of IL-31 in the genesis of generalized pruritus [3]. The importance of the malignant T cells in contributing to increased production of IL-31 is discussed further below.

It has been demonstrated that, in addition to CD4+ T cells, CD8+ T cells may also produce IL-31, but typically at lower levels in comparison to CD4+ T cells [7]. Other cell types that are involved in inflammatory reactions, such as mast cells, eosinophils, basophils, macrophages, and dendritic cells, may also release IL-31 [8, 9]. Moreover, keratinocytes and dermal fibroblasts have also been proposed as possible sources of IL-31 [10]. In the neoplastic setting, malignant CD4+ T cells obtained from patients with CTCL have the capacity to produce IL-31 with levels correlating with the degree of pruritus in advanced disease [7]. In another lymphoproliferative malignancy, follicular B-cell lymphoma, lymph node germinal centers of patients with

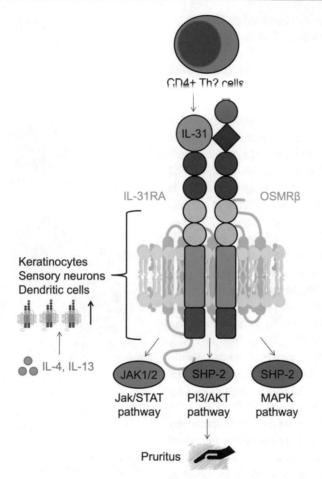

Fig. 8.1 The interaction between IL-31 and its receptor activates signaling pathways that induce pruritus. The IL-31 receptor consists of a heterodimeric receptor composed of IL-31RA and OSMRβ. The receptor is expressed on keratinocytes, dendritic cells, and sensory neurons conducting the itch signal, among other cell types. Binding of IL-31 to its receptor activates the JAK/STAT, PI3K/AKT, and MAPK pathways, through which downstream signaling induces pruritus. The IL-31 receptor is upregulated in response to IL-4 and IL-13. Abbreviations: *CD* cluster of differentiation, *Th2* T helper 2, *IL* interleukin, *IL-31RA* interleukin-31 receptor-alpha, *OSMRβ* oncostatin-M specific receptor subunit beta, *JAK* Janus kinase, *STAT* signal transducer and activator of transcription, *SHP-2* Src homology region 2 domain-containing phosphatase-2, *PI3K* phosphoinositide 3-kinase, *AKT* protein kinase B, *MAPK* mitogen-activated protein kinase

active disease have also been observed to be a source of IL-31 at levels greater than in normal germinal centers [11].

The role of type 2 innate immune cells, referred to as ILC2s, which are critical for the regulation of Th2 T cells, are also likely important in regard to effects on IL-31 production [12]. These cells may be found in the circulation and at higher levels within the skin. Although it is presently uncertain whether ILC2s directly produce IL-31, they do produce IL-4 and IL-13, which play a critical role in the activation, differentiation, and growth of Th2 cells [13]. Moreover, it is these very cytokines, as mentioned below, that may activate various cell types to produce IL-31 and enhance its release.

8.3 IL-31 Interactions with Other Cytokines and Chemokines

IL-31 has significant regulatory interactions with other T-cell-derived cytokines, as well as cytokines produced by innate immune cells. As alluded to previously, IL-31 is produced predominantly by Th2 T cells and has been considered to be a member of the Th2 family of cytokines, with the greatest production by circulating CD4+ Th2 cells. Evidence for the Th2 nature of IL-31 is demonstrated by the clear association of this cytokine with the pathogenesis of Th2-type inflammatory disorders including atopic dermatitis and allergic asthma [14], as well as with CTCL, which tends to have a Th2 bias [7].

ILC2s are important regulators of Th2 T-cell activation and differentiation, and, as such, likely have significant effects on IL-31 production by Th2 cells (Fig. 8.2). ILC2s are regulated by IL-25, which is produced by a variety of hematopoietic cells, cells of the immune system, and epithelial cells, and by IL-33, which can be produced by keratinocytes and other epithelial cells [15]. Furthermore, IL-33 can be rapidly released by keratinocytes in response to cutaneous microbial agents, such as *Staphylococcus aureus*, or by agents that inflict damage to the skin, such as ultraviolet light and scratching [16]. Thus, IL-33 functions as an alarmin with rapid release in response to danger signals leading to innate and adaptive immune responses [17]. In addition, IL-31 may directly induce release of certain cutaneous antimicrobial peptides, including beta-defensins which can further amplify the production of IL-31 through downstream stimulation of IL-33. Both IL-25 and IL-33 are critical in the activation and growth of ILC2s, which in turn are potent producers of IL-4 and IL-13, likely leading to the paracrine enhancement of Th2 cells. Moreover, IL-33 enhances the expression of GATA3, a critical transcription factor for the promotion of Th2 responses leading ultimately to IL-4, IL-5, and IL-13 enhancement [18].

Upon induction of ILC2 cells, IL-4 produced by these cells and, in turn, by Th2 T cells, appears to be an essential factor in promoting IL-31 production by Th2 cells [4]. Moreover,

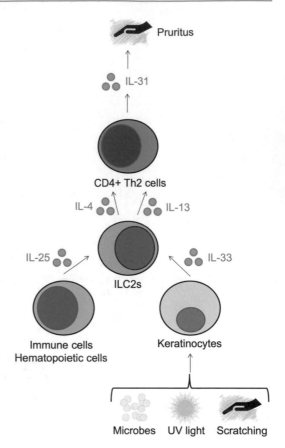

Fig. 8.2 The role of type 2 innate immune cells (ILC2s) in inducing IL-31 production by CD4+ Th2 cells. Type 2 innate immune cells produce IL-4 and IL-13, which play an integral role in the activation, differentiation, and growth of Th2 cells. ILC2s are regulated by IL-25, produced by various hematopoietic and immune cells, and IL-33, produced by keratinocytes in response to cutaneous microbes, ultraviolet light, and scratching. Both IL-25 and IL-33 promote the activation and growth of ILC2s, which in turn produce IL-4 and IL-13, inducing IL-31 production by CD4+ Th2 cells. Abbreviations: *IL* interleukin, *CD* cluster of differentiation, *Th2* T helper 2, *ILC2s* type 2 innate lymphoid cells, *UV* ultraviolet

other cell types under the influence of IL-4, including mast cells, may also demonstrate IL-31 production [19]. Furthermore, there appears to be collaboration between multiple cytokines leading to enhanced IL-31 production. One example is the combined effect of IL-33 with IL-4 to yield increased IL-31 production from mast cells [19]. This effect may be further magnified by IgE on the surface of mast cells and by other local stimuli such as cutaneous beta-defensins, which are released during trauma or infection [20]. In addition, a combination of IL-4 with IL-33 may synergistically enhance IL-31 production by Th2 T cells [21].

Another example of the linked effects of cyto-kines and chemokines to augment IL-31 production, and therefore pruritus within the cutaneous milieu, involves the association between IL-31 and chemokines that are essential for recruitment of CD4+ skin-trafficking T cells, particularly Th2 cells with the capacity to produce additional IL-4 and IL-31. Notably, IL-31 can induce the release of the chemokines CCL17 and CCL22 from keratinocytes and dendritic cells [22]. CCL17 and CCL22 are ligands for the chemokine receptor CCR4, which is expressed on CD4+/CLA+ skin-trafficking T cells and is critical for recruitment of and for homeostasis of T cells in the skin, particularly in response to inflammatory stimuli. Therefore, cutaneous chemokines induced by IL-31 participate in the recruitment of CCR4+/CD4+ skin-trafficking T cells, which are the very cells with the capacity to produce IL-31, subsequently leading to further amplification of the pruritic response. It is also apparent that scratching may release skin defensins from the cutaneous environment leading to enhanced IL-33 production, which can also augment IL-31 production in the local milieu, leading to a repetitive itch-scratch cycle [20].

Another interaction between immune cells and chemokines leading to enhanced potential for pruritus is the observation that local eosinophils may induce the production of CCL17 and CCL22, leading to recruitment of skin-trafficking T cells into the skin [23, 24] (Fig. 8.3). Because the skin-trafficking T cells in this setting are typically Th2 cells that produce IL-5, this cytokine has the potential to amplify the itch response by stimulating the growth of cutaneous eosinophils with further production of CCL17 and CCL22, as well as release of IL-31 from endogenous eosinophils.

Basophils have also been demonstrated to be a source of IL-31 [8]. IL-31 can then induce the release of CCL2 from immune cells which can contribute to basophil recruitment. Basophils have the capacity to release IL-31 in response to IgE-dependent stimulation. Patients with chronic spontaneous urticaria manifest increased skin infiltration with basophils and have increased serum levels of IL-31. Treatment of this condition with omalizumab, which currently binds free IgE, and therefore inhibits IgE binding to both mast cells and basophils, results in clinical improvement with reduced serum levels of IL-31 and thus decreased pruritus [25].

8.4　IL-31 Receptor Expression

The IL-31 receptor is a heterodimeric receptor composed of IL-31 receptor-alpha (IL-31RA) and oncostatin M receptor beta (OSMRβ) [26]. The heterodimeric receptor has been found to be expressed on keratinocytes and on sensory neurons that conduct the itch signal, as well as on multiple other cell types including macrophages and colonic epithelial cells [27]. Binding of IL-31 to its receptor results in the phosphorylation and activation of the JAK/STAT, PI3K/AKT, and MAPK signaling pathways [28]. Blocking of the receptor inhibits downstream signaling within these tissues and can reduce IL-31-mediated pruritus [29]. The receptor appears to be upregulated in response to IL-4 and IL-13, indicating that in clinical conditions manifesting a Th2 bias, there is an increased opportunity to trigger the receptor in response to IL-31, leading to an enhanced state of pruritus. In regard to CTCL, which in advanced forms is clearly associated with a Th2 bias, our collaborative work has demonstrated higher levels of expression of the heterodimeric receptor within the skin of pruritic patients compared to non-pruritic patients with CTCL or within the skin of healthy volunteers [30]. Although the levels of expression were observed to correlate with degree of pruritus among this patient population, they did not correlate with stage of CTCL [30].

Recent evidence from Perrigoue and colleagues involving a murine model of helminth infection has suggested a regulatory circuit between Th2 cytokine expression and expression of the IL-31 receptor [31]. IL-31RA is constitutively expressed in the colon and exposure to *Trichuris* induced the expression of IL-31 in

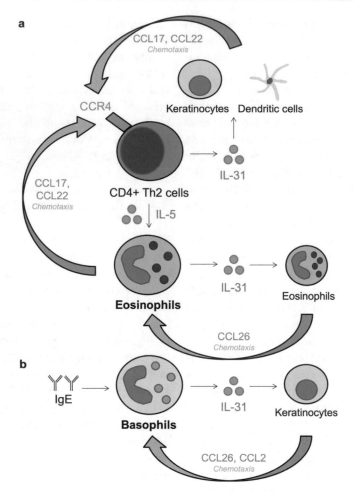

Fig. 8.3 The role of eosinophils and basophils in mediating IL-31 production. (**a**) Eosinophils release IL-31 endogenously in addition to inducing eosinophils to produce CCL17 and CCL22, which are ligands for the chemokine receptor CCR4 on skin-trafficking Th2 cells, recruiting these Th2 cells into the skin. These T cells then produce IL-5, stimulating the growth of eosinophils and further production of IL-31, CCL17, and CCL22. IL-31 also stimulates keratinocytes and dendritic cells to release CCL17 and CCL22. (**b**) Basophils release IL-31 in response to IgE-dependent stimulation. IL-31, in turn, induces the release of CCL2, which promotes basophil recruitment. IL-31 also leads to production of CCL26, a chemotactic factor for both eosinophils and basophils, which then amplify the local inflammatory reaction with further IL-31 production. Abbreviations: *CCR* C-C motif chemokine receptor, *CD* cluster of differentiation, *Th2* T helper 2, *IL* interleukin, *CCL* C-C motif ligand, *Ig* immunoglobulin

CD4+ T cells. In response to *Trichuris* infection, IL-31RA−/− knockout mice exhibited increased Th2 cytokine responses in the mesenteric lymph nodes as well as elevated serum IgE and IgG1 levels compared with wild type IL-31RA expressing mice. Therefore, the expression of the IL-31 receptor appears to mediate a counter-regulatory effect to induce negative feedback on Th2 cytokine release.

8.5 Disease States and IL-31

8.5.1 Atopic Dermatitis

Emerging evidence currently supports the integral role of IL-31 in the pathogenesis of chronic pruritus in numerous inflammatory, autoimmune, and malignant disorders. This concept is particularly apparent among patients with atopic dermatitis and pruritus. Atopic dermatitis is a common chronic inflammatory skin condition present in up to 15% of children and a smaller percentage of adults [32]. Th2 cytokines, including IL-4, IL-5, and IL-13, are markedly elevated in the skin of atopic patients with acute lesions [33, 34]. However, it must be stressed that other cytokines, including increased IL-22, may also play a role in the pathogenesis of atopic dermatitis [35]. Moreover, during the chronic phase of atopic dermatitis, interferon gamma may also play a role in perpetuating inflammation, although some studies have suggested a benefit of treatment with interferon gamma.

In concert with elevated Th2 cytokines, ILC2 cells are also increased in the skin of patients with atopic dermatitis [36]. As mentioned, enhanced production of the epithelial cytokines, IL-25 and IL-33, likely plays a role in the enhanced growth of the ILC2 cells. Increased ILC2 cells with production of Th2 cytokines and other factors leading to Th2 T-cell growth and activation can further magnify the Th2 bias in atopic dermatitis.

The augmented growth of ILC2 cells along with Th2 cells would be expected to further enhance IL-31 production in the skin of atopic patients. It is noteworthy that IL-31 levels in the skin and serum of patients with atopic dermatitis have been observed to be elevated in comparison to normal volunteers [5, 37]. Furthermore, CD4+/CLA+ skin-trafficking T cells isolated from the peripheral blood of patients with atopic dermatitis have been shown to be capable of producing higher levels of IL-31 in comparison to this cellular population isolated from healthy donors or from patients with psoriasis [1].

The importance of the IL-31 receptor in atopic dermatitis has also been demonstrated. IL-31 receptor levels are clearly elevated on the keratinocytes of individuals with atopic dermatitis [38]. Although levels of IL-31 have been elevated in the serum of atopic patients, a consistent relationship between extent of pruritus and absolute levels of IL-31 has not been observed.

A significant clinical experience has been developed using cytokine-blocking agents in an effort to treat the inflammatory and pruritic signs and symptoms among patients with atopic dermatitis. Dupilumab, a frequently used monoclonal antibody that blocks the IL4α receptor, inhibits cytokine signaling by both IL-4 and IL-13. This therapeutic agent, which is especially well tolerated among all age groups, can yield profound improvement in disease manifestations in the majority of treated patients [39]. Importantly, pruritus may be completely eliminated, even among those with the most severe symptoms. Dupilumab, by virtue of blocking IL-4 signaling, may also reduce production of IL-31 (AH Rook; unpublished observations), which then secondarily would lead to reduced pruritus.

Antibodies that neutralize IL-13 have also been used in clinical trials for atopic dermatitis. Both tralokinumab and lebrikizumab have produced significant improvement in the clinical manifestations of atopic dermatitis [40, 41]. It is not clear whether any of the beneficial effects were mediated through an effect on IL-31 production.

Nemolizumab is a humanized monoclonal antibody against the IL-31α receptor [6]. In several clinical trials for atopic dermatitis, this antibody resulted in improvement in dermatitis scores compared to placebo in randomized, blinded studies. Importantly, significant improvements in pruritus were observed and maintained throughout a 52-week period during which patients received the antibody every 4 weeks [42].

8.5.2 Autoimmune and Autoinflammatory Diseases

A number of autoimmune and autoinflammatory conditions affecting the skin can be associated with significant degrees of pruritus (Table 8.1). These conditions include bullous pemphigoid, psoriasis, and dermatomyositis. Among pruritic patients having any of these three clinical conditions, elevated IL-31 levels at sites of cutaneous lesions have been observed [47]. Psoriasis is considered to be a Th17-driven systemic inflammatory disease with variable degrees of skin involvement characterized by scaly

Table 8.1 Current evidence on IL-31 in the skin and serum and its major cellular sources in various dermatologic diseases in humans

Disease	Current evidence	Cellular sources	References
Atopic dermatitis	Increased IL-31 in skin and serum Increased IL-31R in skin	Th2 cells, ILC2s	Cornelissen et al. [38] Meng et al. [37] Nobbe et al. [5] Salimi et al. [36]
Psoriasis	Increased IL-31 in skin	Unknown	Nattkemper et al. [43] Czarnecka-Operacz et al. [44]
Dermatomyositis	Increased IL-31 in skin Increased IL-31R in skin	CD4+ T cells	Kim et al. [45]
Bullous pemphigoid	Increased IL-31 in skin	Eosinophils	Kunsleben et al. [46] Rüdrich et al. [9]
Cutaneous T-cell lymphoma	Increased IL-31 in skin and serum	Th2 cells, CD4 + CD26- cells	Singer et al. [7] Cedeno-Laurent et al. [3]

Abbreviations: *IL-31* interleukin-31, *IL-31R* interleukin-31 receptor, *Th2* T helper 2, *ILC2s* type 2 innate lymphoid cells, *CD* cluster of differentiation

patches and plaques involving the skin surface. Nevertheless, at least one group has detected elevated levels of IL-31 in pruritic psoriatic skin [43]. However, another group failed to observe any correlation between severity of pruritus in psoriasis and serum levels of IL-31 [44]. Thus, levels of IL-31 within the inflamed skin in proximity to sensory nerves that can be activated by IL-31 may be most responsible for increased pruritus.

Dermatomyositis is an autoimmune disorder involving multiple tissues with inflammation predominantly involving skin and muscle. Many patients with this disorder complain of pruritus. In this regard, Kim and colleagues detected increased IL-31 mRNA within active skin lesions compared to clinically uninvolved skin [45]. Moreover, levels of IL-31 mRNA positively correlated with itch score. In addition, increased expression of the IL-31 receptor was detected within active skin lesions on immunofluorescence studies [45]. Examination of immune cells isolated from skin lesions demonstrated a variety of cell types producing IL-31, including CD4+ T cells, CD8+ T cells, and dendritic cells. CD4+ T cells appeared to produce the highest amounts of IL-31 upon activation [45].

Bullous pemphigoid is another autoimmune disorder accompanied by significant pruritus. Bullous pemphigoid is characterized by IgG autoantibody-mediated subepidermal skin blistering with the presence of numerous eosinophils within the skin and within the blister fluid. Active skin lesions are typically associated with marked pruritus. Notably, IL-31 levels have been found to be remarkably elevated within skin blister fluid [9]. Eosinophils within lesions represent the most significant source of IL-31 production [9]. Moreover, eosinophils isolated from the circulation have been observed to be producing increased IL-31. It is also noteworthy that IL-31 may induce the production of CCL26 from eosinophils, serving as a chemotactic agent for eosinophils and basophils that can then further amplify the local inflammatory reaction with additional IL-31 production [46].

8.5.3 Malignancy-Associated Pruritus: A Focus on CTCL

Chronic generalized pruritus is considered to be a sign of possible malignancy. This symptom is particularly associated with lymphomas, both Hodgkin lymphoma as well as non-Hodgkin lymphoma. In that regard, evidence has supported the role of IL-31 in the pathogenesis of pruritus in various lymphomas. The most extensive work has been performed on CTCL, upon which we will focus.

CTCL is a heterogeneous group of non-Hodgkin T-cell lymphomas associated with malignant skin-trafficking lymphocytes [48]. The most common forms of CTCL are mycosis fungoides, characterized by cutaneous macules, patches, plaques and tumors, and Sézary syndrome (SS), characterized by erythroderma, circulating malignant T cells and, often lymphadenopathy [49]. The malignant cells are typically CD4+/CLA+/CCR4+ skin-trafficking T cells exhibiting a Th2 bias [50]. A small percentage of cases exhibit a CD8+ phenotype.

Patients with advanced disease often experience severe pruritus, particularly among patients with erythroderma associated with SS, in which pruritus may be especially unrelenting and intractable [51, 52]. Afflicted patients are often unable to sleep and their ability to concentrate in the workplace can be significantly limited. The skin is typically heavily infiltrated with malignant T cells. These cells, which can be isolated from the skin and blood of erythrodermic patients, exhibit a Th2 profile with increased IL-4 and IL-13 [53], and in many cases, increased IL-5 [54]. Furthermore, Singer and colleagues demonstrated significant perturbations in IL-31 production in both the skin and the blood as well as accompanying significant pruritus in patients with SS [7]. Skin biopsies and peripheral blood cells from pruritic patients demonstrated increased IL-31 mRNA expression (Fig. 8.4). Serum levels of IL-31 protein were also elevated among patients with pruritus in comparison to patients without pruritus or healthy volunteers [7].

It is important to note that during treatment of SS with immunomodulatory therapy—including either interferon alpha or interferon gamma, along with extracorporeal photopheresis and the retinoid bexarotene—IL-31 levels declined in the serum as the degree of pruritus improved [7]. A surprising observation made in several patients was that IL-31 serum levels declined with improvement in pruritus, despite no change in the overall absolute numbers of circulating malignant T cells, which initially suggested that another cell type might be implicated in IL-31 production. Additional studies focused on the isolated, purified malignant T cells, distinguished from non-malignant T cells by either their dominant V beta T-cell receptor, or upon the expression of the well-recognized malignant phenotype

of CD4+/CD26- T cells [7] (Fig. 8.5). When the malignant cells were examined directly, it became clear that these cells were the highest producers of IL-31 among all peripheral blood mononuclear cells isolated from SS patients. Notably, in concert with improvements in the degree of pruritus, reduced IL-31 production on an individual malignant T-cell basis was observed. Thus, although the overall numbers of circulating malignant T cells might not have significantly changed, on an individual malignant cell basis, the cells producing IL-31 declined as did the overall serum levels of IL-31.

In addition to interferons and photopheresis, other therapies for CTCL can ameliorate pruritus and disease manifestations. Two FDA-approved histone deacetylase inhibitors, vorinostat and romidepsin, are notable for their ability to produce significant improvement in generalized pruritus. The studies of Cedeno-Laurent and colleagues demonstrated a reduction in IL-31 mRNA in peripheral blood T cells in parallel with a reduction in pruritus during romidepsin therapy [3] (Fig. 8.6). In another study, culture of patients' peripheral blood lymphocytes with vorinostat also resulted in the inhibited IL-31 production [3] (Fig. 8.7).

Systemic corticosteroids are also known to diminish pruritus among patients with advanced CTCL. To examine whether corticosteroids diminish IL-31 from malignant T cells, Cedeno-Laurent and colleagues cultured patients' lymphocytes with dexamethasone. In the majority of samples tested, dexamethasone significantly inhibited production of IL-31 [3].

Because the malignant CD4+ T cells in the skin and blood of pruritic patients with CTCL are responsible for producing increased levels of IL-31, therapeutic agents that directly target the malignant population would be expected to reduce IL-31 secretion, and, therefore, pruritus. In that regard, mogamulizumab is an anti-CCR4 monoclonal antibody that directly binds to the CCR4 molecule that is typically upregulated on the surface of the malignant population within the peripheral blood [45]. Once bound, the antibody works to eradicate the malignant population through antibody-dependent cell-mediated cytotoxicity.

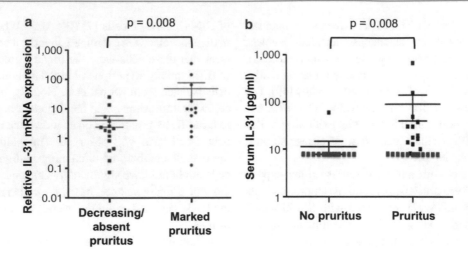

Fig. 8.4 PBMC IL-31 mRNA and serum IL-31 correlates with marked pruritic symptoms in CTCL patients. (**a**) PBMCs were stimulated with PMA/ionomycin. Increased IL-31 mRNA was detected by quantitative RT-PCR from PBMCs of CTCL patients with marked pruritus versus those with decreasing or absent pruritus, $P = 0.006$. Note that all CTCL patient mRNA values were standardized relative to the average values from a selected pool of 13 aged-matched, normal controls. Error bars: mean ± SEM (n: decreasing/absent pruritus = 16; marked pruritus = 13). (**b**) Elevated serum IL-31 levels in CTCL patients with pruritus versus those without pruritus, $P = 0.0083$. Error bars: mean ± SEM (*n*: no pruritus = 14; pruritus = 26). Note that the linear range of the assay was 7.8 pg/mL and that all samples with a value below 7.8 pg/mL or those without any detectable IL-31 were assigned a value of 7.8 pg/mL and are colored in red; thus, non-pruritic patients appear to have a median approaching 7.8 pg. Abbreviations: *PBMC* peripheral blood mononuclear cell, *IL* interleukin, *mRNA* messenger RNA, *CTCL* cutaneous T-cell lymphoma, *PMA* phorbol 12-myristate 13-acetate, *RT-PCR* reverse transcription polymerase chain reaction, *SEM* standard error of the mean, *pg* picograms, *ml* milliliters

Reduction in the circulating population of malignant T cells by mogamulizumab is typically associated with improvement in clinical signs and symptoms of disease, including improvement in pruritus [45]. In fact, mogamulizumab has been shown to reduce the percentage of IL-31-producing T cells [3] (Fig. 8.8).

The ability to reduce the IL-31 production in advanced CTCL is, therefore, a highly favorable goal with the expectation that the remarkably disturbing symptom of pruritus would be reduced. It is anticipated that eliminating chronic scratching behavior would follow the reduction in IL-31 levels, leading to improved skin findings. As mentioned above, reduced scratching would be expected to curb release of IL-33 from keratinocytes, and, therefore, less IL-33-stimulated IL-31 production. Because IL-31 stimulates the release of CCL17 and CCL22, chemokines that recruit Th2 T cells into the skin, reduced IL-31 may also diminish the release of these chemokines considered to be important cofactors for the skin recruitment of the very cells that are IL-31-producing cells. Furthermore, reduced IL-31 would also lead to a decrease in CCL2 and CCL26, leading to diminished recruitment of eosinophils and basophils within the skin. Therefore, therapeutic approaches that directly eliminate IL-31 may inhibit the many synergistic mechanisms that are in play in advanced CTCL, leading to reduced cutaneous inflammation.

8.6 Lessons from Veterinary Medicine

Veterinary medicine has advanced more rapidly than human medicine in the development of newly approved therapeutics for the treatment of pruritic conditions, primarily for allergic dermatitis, in canines. In fact, atopic dermatitis is the most common dermatologic disease treated in canine patients. The discovery that IL-31 also played a role in the canine form of this disease

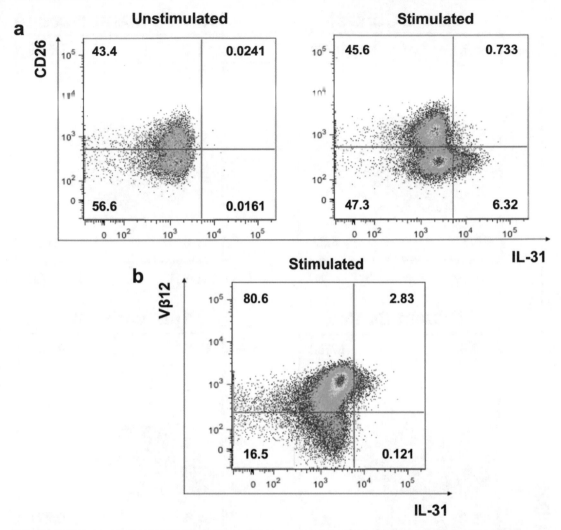

Fig. 8.5 Malignant CD4+ CD26-/CD4+ Vβ + CTCL cells produce IL-31 after stimulation. CD4 T cells, stimulated with PMA/ionomycin, were assessed for intracellular expression of IL-31 by flow cytometry in 15 CTCL patients. (**a**) Representative plots of IL-31 expression from cells without (left) and with (right) stimulation show stimulation of the cells is necessary, leading to production of IL-31 by a small population of T cells. (**b**) Representative plot of IL-31 expression from CD4+ Vβ + CTCL cells. Approximately 6.3% of the stimulated CD4+/CD26- T cells from a highly pruritic patient with stage IVA CTCL were demonstrated to produce IL-31. T cells of non-pruritic patients failed to produce detectable IL-31 (data not shown). Abbreviations: *CD* cluster of differentiation, *CTCL* cutaneous T-cell lymphoma, *IL* interleukin, *PMA* phorbol 12-myristate 13-acetate

led to the initial development of the Janus kinase I (JAK1) inhibitor oclacitinib. Since multiple Th2 cytokines, including IL-4, IL-13, and IL-31 signal through JAK1, the downstream effects of these cytokines are blocked by oclacitinib, yielding additional effects on the underlying pathogenesis of atopy in these patients [55]. When administered to atopic canines, oclacitinib rapidly reduces pruritic behaviors in a significant percentage of patients [55]. While administration of oclacitinib has markedly improved the clinical symptoms of many canines with atopic dermatitis, about one-third of patients do not experience improvement in their pruritus. Furthermore, there is a small risk of adverse effects, including the development of demodicosis in young patients and benign cutaneous histiocytomas in older canines.

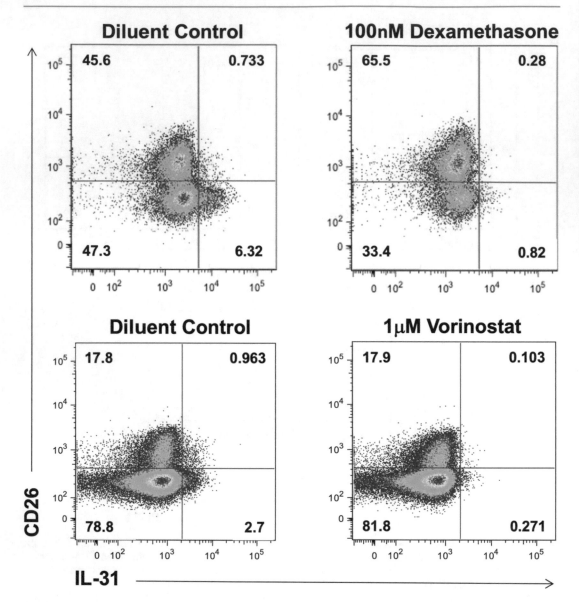

Fig. 8.6 Reduced pruritus and lower IL-31 expression are observed in a CTCL patient after treatment with romidepsin. PBMCs from pruritic advanced-stage CTCL patients were obtained before and after their first infusion of romidepsin. A representative sample was analyzed for IL-31 mRNA by (**a**) quantitative RT-PCR and (**b**) flow cytometry. IL-31 mRNA was no longer detectable at the time pruritus resolved during romidepsin therapy. Abbreviations: *IL-31* interleukin-31, *mRNA* messenger RNA, *CTCL* cutaneous T-cell lymphoma, *RT-PCR* reverse transcription polymerase chain reaction

A second medication, and the first monoclonal antibody approved for use in veterinary medicine, lokivetmab, is an agent that directly targets IL-31. A significant number of canine patients show a rapid decrease in pruritus after receiving lokivetmab [56]. Lokivetmab has very few reported adverse effects and is safe to use in young patients, making it an excellent choice for those who are pruritic with minimal inflammatory skin changes.

The above therapeutics have been demonstrated to have significant benefit for canine atopic dermatitis. As in humans, this inflammatory condition appears to be associated with abnormal Th2 pathophysiology resulting in increased production of IL-13 and IL-31.

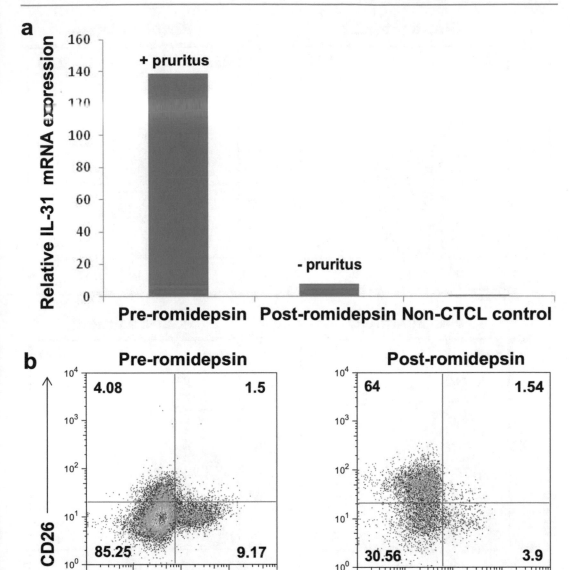

Fig. 8.7 Reduced IL-31 expression with in vitro dexamethasone and vorinostat in samples from CTCL patients. PBMCs from pruritic advanced-stage CTCL patients were treated with 100 nM dexamethasone ($n = 4$) or 1 μM vorinostat ($n = 4$) or their corresponding diluent controls for 12 h. Cells were then stimulated with PMA/ionomycin/brefeldin A for a total of 5 h, stained with fluorophore-conjugated monoclonal antibodies against surface-bound CD3, CD8, CD26, and intracellular IL-31 and were further analyzed by flow cytometry. Representative plots show pre-gated cells on CD3+/CD8-. There was a significant reduction in malignant T cells producing IL-31 after culture with dexamethasone (6.32–0.82%) and vorinostat (2.7–0.271%). Abbreviations: *CD* cluster of differentiation, *IL* interleukin, *nM* nanomoles, *μM* micromoles, *CTCL* cutaneous T-cell lymphoma, *PMA* phorbol 12-myristate 13-acetate

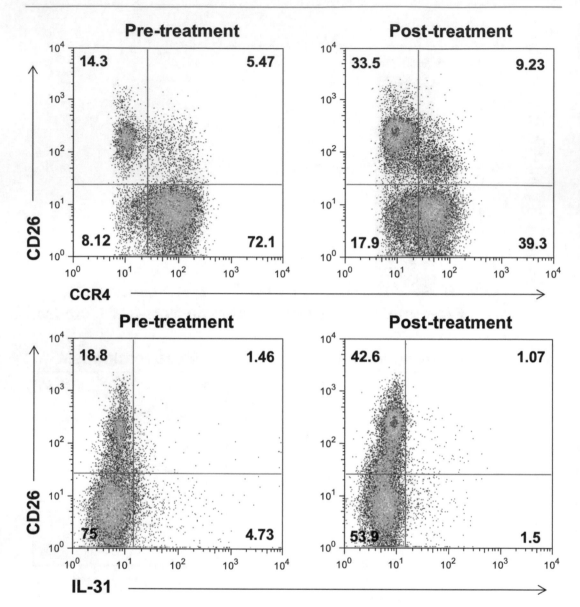

Fig. 8.8 Reduction in the number of CCR4+ T cells via mogamulizumab correlates with decreased numbers of IL-31-expressing T cells. PBMCs from pruritic advanced CTCL samples were obtained before and after the first infusion of mogamulizumab. PBMCs were then stained with fluorophore-conjugated monoclonal antibodies against surface-bound CD3, CD8, CD26, CCR4, and intracellular IL-31 and were further analyzed by flow cytometry. Representative plots show pre-gated cells on CD3+/CD8-. There was a significant reduction in CCR4+ T cells (72.1–39.3%) and in malignant CD26- T cells producing IL-31 (4.73–1.5%) after culture with mogamulizumab. Abbreviations: *CD* cluster of differentiation, *CCR* C-C motif chemokine receptor, *IL* interleukin, *PBMC* peripheral blood mononuclear cell, *CTCL* cutaneous T-cell lymphoma

Therefore, it is unsurprising that lokivetmab, a caninized IL-31 neutralizing antibody, can produce significant reductions in pruritus. As in humans, decreased scratching leads to clinical improvements through reduced inflammatory manifestations in the skin as well as a decreased risk of developing secondary staphylococcal and *Malassezia* skin infections. Importantly, the finding that administration of lokivetmab can delay the onset of an acute atopic flare further highlights the important role of IL-31 in the clinical manifestations of canine atopic dermatitis.

8.7 IL-31 in the Tumor Microenvironment

While immunological consequences of increased IL-31 on the local milieu within lesions of CTCL, in other malignant conditions, have not been fully elucidated, recent evidence suggests that raised levels of this cytokine may confound the ability to generate adequate antitumor immunity. One essential ingredient of antitumor immunity is the generation of interferon gamma by Th1 cells leading to the recruitment of cytotoxic T cells and other critical antitumor cells. This effect is mediated via the release of interferon gamma-induced chemokines. In this regard, the studies of Feld and colleagues [57] have demonstrated that culture of microvascular endothelial cells with interferon gamma upregulates the expression of the IL-31 receptor components on the endothelial cells. The subsequent triggering of the IL-31 receptor by this cytokine can then lead to reduced production of the important interferon gamma-induced chemokine, Mig, and possibly other proteins critical for recruitment of Th1-mediated antitumor cellular elements [57]. Thus, paradoxically, within the tumor milieu associated with high levels of IL-31, this cytokine may actually lead to a negative feedback interplay with interferon gamma on the ability to generate an immune response against the cancer. The net effect of the cytokine interplay on the host immune response may depend upon the relative levels of the Th1 cytokine interferon gamma and the Th2 cytokine IL-31.

Reducing IL-31 levels within CTCL skin lesions may produce beneficial effects that could be quite evident. It would be anticipated that the initial effect of reducing IL-31 would be the elimination of chronic scratching behavior leading to improved skin findings on clinical examination. As mentioned above, reduced scratching would be expected to lead to reduced trauma to keratinocytes and thus a decline in the release of IL-33 from keratinocytes. This effect almost certainly would diminish IL-33-stimulated IL-31 production. Because IL-31 stimulates the release of CCL17 and CCL22, chemokines that recruit Th2 T cells into the skin, reducing IL-31 might also diminish the release of these chemokines, which are considered important cofactors for the recruitment of the very T cells, Th2 cells, that are IL-31-producing cells, into the skin. Th2 cells are also responsible for the production of a multitude of cytokines, including IL-4, IL-5, and IL-13, that may play a critical role in enabling the malignant T cells to circumvent Th-1-mediated host antitumor immunity by virtue of suppression of interferon gamma by these cytokines. Furthermore, a reduction in IL-31 would also lead to decreased induction of CCL2 and CCL26, leading to decreased recruitment of eosinophils and basophils into the skin. These cells also contribute proinflammatory effects with further release of IL-31. Therefore, it would be beneficial to diminish their recruitment into the skin in the setting of CTCL. Thus, therapeutic approaches that directly eliminate IL-31 could directly ameliorate the many synergistic mechanisms that are in play in advanced CTCL that hamper the ability of the host immune response to generate an adequate Th1-driven antitumor cell-mediated immune response.

References

1. Dillon SR, Sprecher C, Hammond A et al (2004) Interleukin 31, a cytokine produced by activated T cells, induces dermatitis in mice [published correction appears in Nat Immunol. 2005 Jan;6(1):114]. Nat Immunol 5(7):752–760
2. Bilsborough J, Leung DY, Maurer M et al (2006) IL-31 is associated with cutaneous lymphocyte antigen-positive skin homing T cells in patients with atopic dermatitis [published correction appears in J Allergy Clin Immunol. 2006;117(5):1124 . Boguniewcz, Mark [corrected to Boguniewicz, Mark]]. J Allergy Clin Immunol 117(2):418–425
3. Cedeno-Laurent F, Singer EM, Wysocka M et al (2015) Improved pruritus correlates with lower levels of IL-31 in CTCL patients under different therapeutic modalities. Clin Immunol 158(1):1–7
4. Stott B, Lavender P, Lehmann S, Pennino D, Durham S, Schmidt-Weber CB (2013) Human IL-31 is induced by IL-4 and promotes TH2-driven inflammation. J Allergy Clin Immunol 132(2):446–54.e5
5. Nobbe S, Dziunycz P, Mühleisen B et al (2012) IL-31 expression by inflammatory cells is preferentially elevated in atopic dermatitis. Acta Derm Venereol 92(1):24–28

6. Ruzicka T, Mihara R (2017) Anti-Interleukin-31 receptor A antibody for atopic dermatitis. N Engl J Med 376(21):2093

7. Singer EM, Shin DB, Nattkemper LA et al (2013) IL-31 is produced by the malignant T-cell population in cutaneous T-cell lymphoma and correlates with CTCL pruritus. J Invest Dermatol 133(12):2783–2785

8. Raap U, Gehring M, Kleiner S et al (2017) Human basophils are a source of—and are differentially activated by—IL-31. Clin Exp Allergy 47(4):499–508

9. Rüdrich U, Gehring M, Papakonstantinou E et al (2018) Eosinophils are a major source of Interleukin-31 in bullous pemphigoid. Acta Derm Venereol 98(8):766–771

10. Bağci IS, Ruzicka T (2018) IL-31: a new key player in dermatology and beyond. J Allergy Clin Immunol 141(3):858–866

11. Ferretti E, Tripodo C, Pagnan G et al (2015) The interleukin (IL)-31/IL-31R axis contributes to tumor growth in human follicular lymphoma. Leukemia 29(4):958–967

12. Moro K, Yamada T, Tanabe M et al (2010) Innate production of T(H)2 cytokines by adipose tissue-associated c-Kit(+)Sca-1(+) lymphoid cells. Nature 463(7280):540–544

13. Monticelli LA, Sonnenberg GF, Abt MC et al (2011) Innate lymphoid cells promote lung-tissue homeostasis after infection with influenza virus. Nat Immunol 12(11):1045–1054

14. Kim J, Kim BE, Leung DYM (2019) Pathophysiology of atopic dermatitis: clinical implications. Allergy Asthma Proc 40(2):84–92

15. Huang Y, Guo L, Qiu J et al (2015) IL-25-responsive, lineage-negative KLRG1(hi) cells are multipotential 'inflammatory' type 2 innate lymphoid cells. Nat Immunol 16(2):161–169

16. Cayrol C, Girard JP (2018) Interleukin-33 (IL-33): a nuclear cytokine from the IL-1 family. Immunol Rev 281(1):154–168

17. Liew FY, Girard JP, Turnquist HR (2016) Interleukin-33 in health and disease. Nat Rev Immunol 16(11):676–689

18. Schmitz J, Owyang A, Oldham E et al (2005) IL-33, an interleukin-1-like cytokine that signals via the IL-1 receptor-related protein ST2 and induces T helper type 2-associated cytokines. Immunity 23(5):479–490

19. Petra AI, Tsilioni I, Taracanova A, Katsarou-Katsari A, Theoharides TC (2018) Interleukin 33 and interleukin 4 regulate interleukin 31 gene expression and secretion from human laboratory of allergic diseases 2 mast cells stimulated by substance P and/or immunoglobulin E. Allergy Asthma Proc 39(2):153–160

20. Cornelissen C, Brans R, Czaja K et al (2011) Ultraviolet B radiation and reactive oxygen species modulate interleukin-31 expression in T lymphocytes, monocytes and dendritic cells. Br J Dermatol 165(5):966–975

21. Maier E, Werner D, Duschl A, Bohle B, Horejs-Hoeck J (2014) Human Th2 but not Th9 cells release IL-31 in a STAT6/NF-κB-dependent way. J Immunol 193(2):645–654

22. Miake S, Tsuji G, Takemura M et al (2019) IL-4 augments IL-31/IL-31 receptor alpha interaction leading to enhanced Ccl 17 and Ccl 22 production in dendritic cells: implications for atopic dermatitis. Int J Mol Sci 20(16):4053

23. Sugaya M (2015) Chemokines and skin diseases. Arch Immunol Ther Exp 63(2):109–115

24. Liu LY, Bates ME, Jarjour NN, Busse WW, Bertics PJ, Kelly EA (2007) Generation of Th1 and Th2 chemokines by human eosinophils: evidence for a critical role of TNF-alpha. J Immunol 179(7):4840–4848

25. Altrichter S, Hawro T, Hänel K et al (2016) Successful omalizumab treatment in chronic spontaneous urticaria is associated with lowering of serum IL-31 levels. J Eur Acad Dermatol Venereol 30(3):454–455

26. Zhang Q, Putheti P, Zhou Q, Liu Q, Gao W (2008) Structures and biological functions of IL-31 and IL-31 receptors. Cytokine Growth Factor Rev 19(5–6):347–356

27. Ferretti E, Corcione A, Pistoia V (2017) The IL-31/IL-31 receptor axis: general features and role in tumor microenvironment. J Leukoc Biol 102(3):711–717

28. Kasraie S, Niebuhr M, Werfel T (2013) Interleukin (IL)-31 activates signal transducer and activator of transcription (STAT)-1, STAT-5 and extracellular signal-regulated kinase 1/2 and down-regulates IL-12p40 production in activated human macrophages. Allergy 68(6):739–747

29. Kasutani K, Fujii E, Ohyama S et al (2014) Anti-IL-31 receptor antibody is shown to be a potential therapeutic option for treating itch and dermatitis in mice. Br J Pharmacol 171(22):5049–5058

30. Nattkemper LA, Martinez-Escala ME, Gelman AB et al (2016) Cutaneous T-cell lymphoma and pruritus: the expression of IL-31 and its receptors in the skin. Acta Derm Venereol 96(7):894–898

31. Perrigoue JG, Zaph C, Guild K, Du Y, Artis D (2009) IL-31-IL-31R interactions limit the magnitude of Th2 cytokine-dependent immunity and inflammation following intestinal helminth infection. J Immunol 182(10):6088–6094

32. Silverberg JI (2017) Public health burden and epidemiology of atopic dermatitis. Dermatol Clin 35(3):283–289

33. Hussein YM, Shalaby SM, Nassar A, Alzahrani SS, Alharbi AS, Nouh M (2014) Association between genes encoding components of the IL-4/IL-4 receptor pathway and dermatitis in children. Gene 545(2):276–281

34. Esaki H, Brunner PM, Renert-Yuval Y et al (2016) Early-onset pediatric atopic dermatitis is T_H2 but also T_H17 polarized in skin. J Allergy Clin Immunol 138(6):1639–1651

35. Nograles KE, Zaba LC, Shemer A et al (2009) IL-22-producing "T22" T cells account for upregulated IL-22 in atopic dermatitis despite reduced IL-17-producing TH17 T cells. J Allergy Clin Immunol 123(6):1244–52.e2

36. Salimi M, Barlow JL, Saunders SP et al (2013) A role for IL-25 and IL-33-driven type-2 innate lymphoid cells in atopic dermatitis. J Exp Med 210(13):2939–2950

37. Meng J, Moriyama M, Feld M et al (2018) New mechanism underlying IL-31-induced atopic dermatitis. J Allergy Clin Immunol 141(5):1677–1689.e8

38. Cornelissen C, Marquardt Y, Czaja K et al (2012) IL-31 regulates differentiation and filaggrin expression in human organotypic skin models. J Allergy Clin Immunol 129(2):426–433.e4338

39. Simpson EL, Bieber T, Guttman-Yassky E et al (2016) Two phase 3 trials of dupilumab versus placebo in atopic dermatitis. N Engl J Med 375(24):2335–2348

40. Popovic B, Breed J, Rees DG et al (2017) Structural characterisation reveals mechanism of IL-13-neutralising monoclonal antibody Tralokinumab as inhibition of binding to IL-13Rα1 and IL-13Rα2. J Mol Biol 429(2):208–219

41. Kasaian MT, Raible D, Marquette K et al (2011) IL-13 antibodies influence IL-13 clearance in humans by modulating scavenger activity of IL-13Rα2. J Immunol 187(1):561–569

42. Kabashima K, Furue M, Hanifin JM et al (2018) Nemolizumab in patients with moderate-to-severe atopic dermatitis: randomized, phase II, long-term extension study. J Allergy Clin Immunol 142(4):1121–1130.e7

43. Nattkemper LA, Tey HL, Valdes-Rodriguez R et al (2018) The genetics of chronic itch: gene expression in the skin of patients with atopic dermatitis and psoriasis with severe itch. J Invest Dermatol 138(6):1311–1317

44. Czarnecka-Operacz M, Polańska A, Klimańska M et al (2015) Itching sensation in psoriatic patients and its relation to body mass index and IL-17 and IL-31 concentrations. Postepy Dermatol Alergol 32(6):426–430

45. Kim HJ, Zeidi M, Bonciani D et al (2018) Itch in dermatomyositis: the role of increased skin interleukin-31. Br J Dermatol 179(3):669–678

46. Kunsleben N, Rüdrich U, Gehring M, Novak N, Kapp A, Raap U (2015) IL-31 induces chemotaxis, calcium mobilization, release of reactive oxygen species, and CCL26 in eosinophils, which are capable to release IL-31. J Invest Dermatol 135(7):1908–1911

47. Gibbs BF, Patsinakidis N, Raap U (2019) Role of the pruritic cytokine IL-31 in autoimmune skin diseases. Front Immunol 10:1383

48. Diamandidou E, Cohen PR, Kurzrock R (1996) Mycosis fungoides and Sezary syndrome. Blood 88(7):2385–2409

49. Campbell JJ, Clark RA, Watanabe R, Kupper TS (2010) Sezary syndrome and mycosis fungoides arise from distinct T-cell subsets: a biologic rationale for their distinct clinical behaviors. Blood 116(5):767–771

50. Guenova E, Watanabe R, Teague JE et al (2013) TH2 cytokines from malignant cells suppress TH1 responses and enforce a global TH2 bias in leukemic cutaneous T-cell lymphoma. Clin Cancer Res 19(14):3755–3763

51. Sampogna F, Frontani M, Baliva G et al (2009) Quality of life and psychological distress in patients with cutaneous lymphoma. Br J Dermatol 160(4):815–822

52. Lewis DJ, Huang S, Duvic M (2018) Inflammatory cytokines and peripheral mediators in the pathophysiology of pruritus in cutaneous T-cell lymphoma. J Eur Acad Dermatol Venereol 32(10):1652–1656

53. Vowels BR, Cassin M, Vonderheid EC, Rook AH (1992) Aberrant cytokine production by Sezary syndrome patients: cytokine secretion pattern resembles murine Th2 cells. J Invest Dermatol 99(1):90–94

54. Suchin KR, Cassin M, Gottleib SL et al (2001) Increased interleukin 5 production in eosinophilic Sézary syndrome: regulation by interferon alfa and interleukin 12. J Am Acad Dermatol 44(1):28–32

55. Cosgrove SB, Wren JA, Cleaver DM et al (2013) Efficacy and safety of oclacitinib for the control of pruritus and associated skin lesions in dogs with canine allergic dermatitis. Vet Dermatol 24(5):479–e114

56. Souza CP, Rosychuk RAW, Contreras ET, Schissler JR, Simpson AC (2018) A retrospective analysis of the use of lokivetmab in the management of allergic pruritus in a referral population of 135 dogs in the western USA. Vet Dermatol 29(6):489–e164

57. Feld M, Shpacovitch VM, Fastrich M, Cevikbas F, Steinhoff M (2010) Interferon-γ induces upregulation and activation of the interleukin-31 receptor in human dermal microvascular endothelial cells. Exp Dermatol 19(10):921–923

Index

A
Adalimumab, 86
Adaptive immune system, 90
Adenoviruses (AdV)
  E1 and E3 genes, 72
  first-generation oAdV, 72
  IL-12-encoding, 72
  oAdV, 72
  primary tumor growth, 72
  radiation, 73
  second-generation oAdVs, 73
  trials, 73
  vectors, 73
Adjuvanticity, 72
Androgen-insensitive cancer cells, 5
Androgen receptor (AR), 1
Angiogenesis, 93–95
  IL-24 role, 102, 103
Antigenicity, 72
Anti-IL-6 antibodies, 5
Antitumor immunity, 104
Apoptosis, 100, 101
  drug-induced, 32
  IL-7, 28
  inhibition, 23
  NKT cells, 24
  rapamycin-induced, 23
Apoptotic-mediated cell death, 101
Atopic dermatitis
  chronic inflammatory skin condition, 117
  cytokine-blocking agents, 117
  dermatologic disease, 120
  IL-31 levels, 117
  IL-31 receptor, 117
  ILC2 cells, 117
  nemolizumab, 117
  role, IL-22, 117
  role, IL-31, 112, 118
  Th2 bias, 112
  therapeutics, 122
  tralokinumab and lebrikizumab, 117

B
Basic leucine zipper ATF-like transcription factor
  (BATF), 95
B-cell acute lymphoblastic leukemia (B-ALL), 33
B-cell precursor acute lymphoblastic leukemia (BCP-
  ALL), 14
Breast cancer
  IL-22 role, 85, 86
Briakinumab, 96
Bullous pemphigoid, 118

C
Cancer-associated fibroblasts (CAFs), 11
Cancer stem cells (CSCs), 52
  IL-24 role, 103
Cancer therapy, 69, 70
Carcinoma
  EAC, 34
  HCC, 38
  hepatocellular, 11
  IL-7, cancer cell migration, 27
  RCC, 37
CD4+ naïve T cells, 55
CD4+ T cells, 53, 54, 57
CD8+ cytotoxic T cells, 55
Cell-based immunotherapy, 55
Cell penetrating domains (CPDs), 56
Checkpoint (CP) blockade, 69, 70, 72, 75
Checkpoint inhibitors (CPIs), 69
Chemokine receptor-4 (CXCR-4), 103
Chemokines, 2
  CCR 7, 57
  DCs and T cells, 57
  downmodulation, 57
  gamma-induced, 125
  and IL-31, 115, 120
  interleukins, 53
Chimeric antigen receptor (CAR) T-cell
  therapies, 56, 61
Chronic generalized pruritus, 118

Chronic myeloid leukemia (CML), 34
Classic Hodgkin's lymphoma (cHL), 34
Colorectal cancer (CC/CRC)
    BATF, 95
    IL-22 role, 84–85
    IL-23 expression, 95
    immunotherapy, 95
Cutaneous T-cell lymphoma (CTCL), 33
    IL-31 production, 120
    leukemic forms, 112
    malignant CD4+ T cells, 119
    non-Hodgkin T-cell lymphomas, 118
    skin lesions, 125
    systemic corticosteroids, 119
    Th2 bias, 115
    vorinostat and romidepsin, 119
Cyclobutane pyrimidine dimer (CPD), 93
Cytokines, 100, 104
    IL-12 family, 90
    IL-23 (*see* Interleukin-23 (IL-23))
    TME, 90
    in tumorigenesis, 90

D
Dendritic cells (DCs), 52
    adaptive immune response, 55
    conventional/myeloid, 53
    immunotherapy, 56
    plasmacytoid, 53
    and T-cell-based cellular immunotherapies, 53
    vaccine trials, 55
Dermatomyositis, 111, 117, 118
7,12-Dimethylbenz (a) anthracene (DMBA),
    92, 93
Downstream signaling, 90
Drug resistance, 14
    phosphorylation, YB-1, 23
    PI3K/Akt pathway, 22
    YB-1, 33
Dupilumab, 117

E
Ehrlich's ascites carcinoma (EAC), 34
Elimination phase, 52
Endogenous inhibitors, IL-6 signaling, 5
Epidermal growth factor receptor (EGFR), 102
Epithelial-mesenchymal transition (EMT), 4, 22, 26,
    53, 54
    E-cadherin expression, 26
    EMT inducer, 26
    IL-7/IL-7R, 26
    IL-7δ5-mediated EMT, 38
    markers, 26
    Pim-1/NFAT2 activation, 26
    potential role, IL-7, 26
    in prostate cancer, 35
Esophageal cancer
    IL-23 expression, 95, 96

Etanercept, 86
Extracellular matrix (ECM), 90

F
Focal adhesion kinase (FAK), 103

G
Galiellalactone, 5
Gastric cancer (GC)
    IL-22 role, 85
Glycosylation, 100

H
Healthy immune system, 52
Human hepatocellular carcinoma
    (HCC), 83

I
IL-7/IL-7R signaling
    and autophagy, 27
    B-ALL, 33
    bladder cancer, 37
    breast cancer, 38
    cancer angio- and lymphangiogenesis, 27
    cancer immunosuppression, 34
    cancer invasion and metastasis, 26–27
    cell growth and metabolism, 25–26
    cell proliferation and cell cycle progression,
        24–25
    cell survival, 23–24
    in cHL, 34
    CML, 34
    CRC, 36
    and effectiveness of anticancer therapy, 28
    esophageal cancer, 35
    fibrosarcoma, 35
    gastric cancers, 36
    gnecological cancers, 37
    HCC, 38
    hematological malignancies, 30
        CTCL, 33
        T-ALL, 31–33
    immunomodulatory role in TME, 28–30
    JAK/STAT, 14–20
    lung cancer, 36
    melanoma cells, 37
    MM, 33, 34
    PI3K/Akt, 20–22
    prostate cancer, 35
    Pyk2 phosphorylation, 23
    RCC, 37
    signaling cascades, 15
    Src kinases p56[lck] and p59[fyn], 23
IL-12 receptor (IL-12R), 68
IL-20 receptor (IL-20R), 101–103
IL-22 binding protein (IL-22BP), 86

IL-22 receptor (IL-22R), 102
IL-31 receptor
    in atopic dermatitis, 117
    heterodimeric receptor, 115
    IL-31RA, 115
    pruritic atopic patients, 112
Immune checkpoints, 59
Immune healing, 72
Immunosuppressive bioactive compounds, 53
Immunosurveillance, 30, 34, 35, 52
Immunotherapies, 53
    with IL-7 (see Interleukin-7 (IL-7))
    interleukin potential, 10
Immunotherapy, 56
Inducible nitric oxide synthase (iNOS), 101
Infliximab, 86
Innate immune system, 90
Innate lymphocytes, 82
Interferon gamma (IFNγ), 59, 61, 90
Interleukin-4 (IL-4)
    CD4+ skin-trafficking T cells, 115
    dupilumab, 117
    ILC2 cells, 114
    and IL-13, 112, 113
    Th2 cytokines, 117
    Th2 responses, 114
Interleukin-6 (IL-6)
    and androgen responsiveness in PCa, 4
    anti-IL-6 antibodies, 5
    binding, 6
    endogenous inhibitors, 5
    and PCa therapy, 5
    proliferation, 3, 5
    sensitivity during PCa progression, 5
    signaling pathways in PCa, 2–4
Interleukin-7 (IL-7)
    BAC transgenic mice, 11
    early lymphocyte development, 10
    functionality, 10
    IL-7/IL-7R signaling (see IL-7/IL-7R
        signaling)
    IL-7R, 11, 12
    immunomodulatory role, 31
    interaction with ECM, 13
    and interleukin potential, 10
    25 kD glycoprotein, 10
    in melanoma cells, 11
    potential sources, 10
    synthesis, 11
    for T cells, 10
    tissue-based interleukin, 10
Interleukin-7 receptor (IL-7R), 11, 12
    See also IL-7/IL-7R signaling
Interleukin-10 (IL-10), 100, 104
    as cytokine synthesis inhibitory factor, 53
    on DC-mediated immune responses,
        55–56
    NF-κB signaling pathway, 60–61
    in OC microenvironment, 53–55
    production of pro-inflammatory cytokines, 59

Interleukin-12 (IL-12)
    activated macrophages, 68
    AdV, 72
    in antitumor immunity, 90
    approaches, 68
    Clinicaltrials.gov website, 68
    cytokine family, 90
    discovery, 68
    IL-12R, 68
    innate and adaptive responses, 68
    NDV, 74
    NK and CD8+ T-cell activation, 68
    oHSV, 69–72
    oMeV, 74
    OVs, 69
    p70, 68
    potent anticancer effects, 68
    proangiogenic cytokines, 68
    as recombinant protein, 68
Interleukin-13 (IL-13), 112, 113
    cytokine signaling, 117
    ILC2s, 114
    Th2 cytokines, 117, 121
Interleukin-19 (IL-19), 100
Interleukin-22 (IL-22)
    anti-IL-22 drugs, 86
    breast cancer, 85, 86
    downstream signaling, 82, 86
    drug discovery, 86–87
    endogenous, 82
    in GC, 85
    IL-10 cytokine family, 81
    and immunocytes, in TME, 82
    leukemia and lymphoma, 86
    liver cancer, 83, 84
    lung cancer, 85
    and malignant cells, 82, 83
    oncogenic roles, 83
    in PC, 85
    R2 receptors, 82
    signaling pathways, 82
    as tumor-promoting cytokine, 82
Interleukin-23 (IL-23)
    breast cancer, 93, 94
    in carcinogenesis, 93
    in CRC, 95
    esophageal cancer, 95, 96
    heterodimeric cytokine, 91
    innate and adaptive immune responses, 91
    JAK-STAT pathway, 92
    keratinocyte carcinomas, 93
    melanoma, 93
    MM cells, 94
    non-immunological effects, 92–93
    in pediatric B-ALL, 95
    in therapeutics
        anti IL-12/23p19, 96
        anti IL-12/23p40, 96
    on tumor growth, 95
    on tumorigenesis, 94

Interleukin-24 (IL-24)
   as activator, immune cell function, 104–105
   antiangiogenic activity, 105
   antitumor immunity, 104
   as bio-therapeutic for cancer, 105
   CD4+ proliferation, 105
   cDNA, 100
   combinatorial therapy, 102
   delivery, 100
   endogenous IL-24 expression, 105
   exogenous expression, 101
   features, 100
   as inhibitor
      cell invasion and metastasis, 103
        CSCs, 103
      tumor angiogenesis, 102–103
   and iNOs, 101
   as MDA-7, 100
   metastasis, 103
   as monotherapy, 102
   mRNA expression, 105
   multifunctional properties, 105
   phosphorylation, 100, 101
   protein expression, 100
   protein sequence, 100
   protein-mediated cell death, 101–102
   TSG, 100
   ubiquitination, 100
   viral and non-viral vectors, 101
Interleukin-31 (IL-31)
   atopic dermatitis, 116–117
   autoimmune and autoinflammatory conditions, 117–118
   basophils, 115
   CCR4+ T cells via mogamulizumab, 124
   CD4+ skin-trafficking T cells, 115
   cellular production, 112–113
   CTCL, 118–120
   description, 112
   discovery, 120
   eosinophils and basophils, 116
   etiology of pruritus, 112
   functional cloning approach, 112
   with ILC2s, 114–115
   receptor expression, 115–116
   in TME, 125
   veterinary medicine, 120–124
Interleukin-33 (IL-33)
   functions, 114
   GATA3 expression, 114
   in ILC2s, 114, 117
   with IL-4, 114
   by keratinocytes, 114
   skin defensins, 115

J
Janus kinases, 2
   JAK1, 54
Janus kinase–signal transducers and activators of
       transcription (JAK-STAT), 90
   JAK-STAT3 pathway, 55

K
Keratinocyte carcinomas, 93

L
Leukemia
   ALL, 12
   B-ALL, 33
   BCP-ALL, 14
   CML, 28
   IL-22 role, 86
   T-ALL, 31
Liver cancer
   IL-22 role, 83, 84
Lokivetmab, 122, 124
Lung cancer
   IL-22 role, 85
Lymphangiogenesis, 22, 27
Lymphomas
   CD4+ T cells, 30
   in cHL, 34
   classic Hodgkin's, 30
   CTCL, 11, 33
   and human leukemias, 32
   IL-7 and IL-7R expression, 11
   IL-22 role, 86
   NFAT2 overexpression, 18

M
Maraba virus (MRB), 75
Matrix metallopeptidase 9 (MMP9), 93, 96, 97
Measles, 74
Melanocytic nevi, 93
Melanoma differentiation-associated-gene 7 (MDA-7)
   administration, 101
   differentiation and cell cycle arrest, 100
   IL-24 (*see* Interleukin-24 (IL-24))
   multifunctional properties, 106
Mesothelin CAR T cells, 61
Metastasis
   cancer advancement, 36
   cancer invasion, 13
   CTLs' activation, 30
   expression of IL-7/IL-7R, 35
   IL-7 and CD127 expression, 36
   osteolytic, 34
   tumor progression, 26
Mitogen-activated protein kinase (MAPK), 3, 4
Monoclonal antibody, 122
Monotherapy, 73, 75
Multiple myeloma (MM), 33, 34, 94
Mycosis fungoides, 119

N
Natural killer (NK) cells, 68
Natural killer T (NKT) cells, 82
Nemolizumab, 117
Neoangiogenesis, 92
Neuroendocrine cells, 3

Neuroendocrine phenotype, 3
Nuclear factor-kappa B (NF-κB), 2
    anti-apoptotic and immunomodulatory functions, 60
    DC immunotherapeutic protocols, 60
    description, 60
    immunosuppressive cytokines, 61
    inhibitor IκB in melanoma, 60
    OIV-induced microenvironment, 60

O
Oclacitinib, 121
Onco-immunotherapeutic viruses (OIVs)
    ability to immune suppression, 69
    administration, 74
    clinical trials, 73
    IL-12 cassette, 70
    IL-12-expressing, 69
    immunotherapeutic effects, 69
    immunotherapy, 69
    in murine models, 75
Oncolytic HSV (oHSV)
    antitumor immunity and IL-12, 72
    development, IL-12, 70
    G207, 69
    M002, IL-12 gene, 71
    "safe-and-robust", 71
    tropism, 71
Oncolytic immunotherapy, 69
Oncolytic measles virus (oMeV), 74
Oncolytic viruses (OVs)
    immunotherapy, 69
    intrinsic properties, 69
    lysis, 69
    necroptosis, 69
Ovarian cancer (OC)
    cell-based immunotherapy, 55
    DC- and T-cell-based cellular immunotherapies, 53
    human OC cell line SKOV-3, 57
    IL-10 expression, 52
    immunosuppression, 52
    interleukins role, 53
    mortality, 52
    stem cells, 53
    TGF-β, 52

P
Pancreatic cancer (PC)
    IL-22 role, 85
Pediatric B-acute lymphocytic leukemia (B-ALL), 95
Pericytes, 93
Peritoneal ascitic fluid, 52
Post-translational modification (PTM), 100
Proinflammatory cytokines, 68–70
Prostate cancer (PCa)
    antiandrogen therapy, 2
    anti-STAT3 therapy, 5
    AR, 1
    cellular response to IL-6, 5
    clinical studies, 1, 5

CNTO328 administration, 5
    IL-6 activation, signaling pathways, 2–4
    IL-6 expression, 2
    LNCaP cells, 5
    methodological approaches, 2
    NF kappa B, 2
    therapy, 1, 5
Protein inhibitors of activated STAT (PIAS), 5
Psoriasis, 117
Psoriasis therapy, 96

R
Rapamycin, 32
Receptor activator of nuclear factor kappa-B
        (RANKL), 92
Recombinant adeno-associated viral (rAAV)
        vectors, 56, 57

S
Sézary syndrome (SS), 119
Signal transducer and activator of transcription
        (STAT), 11
    anti-STAT3 therapy, 5
    in PCa, 3
    SOCS, 5
    STAT3, 2, 54
    STAT5, 2
    STAT1/3/5, 82
STAT3 signaling, 90
Stem cells, 4
    OCs, 53
Suppressors of cytokine signaling (SOCS), 5

T
Talimogene laherparepvec (T-Vec), 69
T-cell acute lymphoblastic leukemia (T-ALL), 31–33
T cells, 82, 83
T-helper type 2 (Th2) cells
    CD4+ Th2 cells, 114
    as ILC2s, 113, 114
    IL-31, 112
    Th2 cytokines, 117, 121
Toll-like receptor (TLR) ligands, 90, 91
Transforming growth factor (TGF)-β, 11
Tumor-associated macrophages (TAMs), 94
Tumor infiltrating lymphocytes (TILs), 104
Tumor metastasis, 103
Tumor microenvironment (TME)
    CAFs in Hodgkin's lymphoma, 11
    cytokines, 90
    IL-12 (see Interleukin 12 (IL-12))
    IL-22 and immunocytes, 82
    immunomodulatory role, IL-7/IL-7R, 28–30
    malignant cells, 90
    NFAT2-expressing cells, 18
    tumor pathogenesis, 90
Tumor necrosis factor alpha (TNF-α) antibody, 86
Tumor progression, 2, 4, 5

Tumor suppressor gene (TSG), 100, 101
Type 2 innate immune cells (ILC2s), 113, 114,
        117, 118

U
Ubiquitination, 100
Ustekinumab, 96

V
Vaccine strain of MeV (MeVac), 74
Vascular endothelial growth factor (VEGF), 102
Veterinary medicine, 120

Y
Y-box-binding protein 1 (YB-1), 23, 33

Printed in the United States
by Baker & Taylor Publisher Services